Catalytic Processes for The Valorisation of Biomass Derived Molecules

Catalytic Processes for The Valorisation of Biomass Derived Molecules

Special Issue Editors

Francesco Mauriello
Signorino Galvagno
Claudia Espro

MDPI • Basel • Beijing • Wuhan • Barcelona • Belgrade

MDPI

Special Issue Editors
Francesco Mauriello
Università Mediterranea di Reggio Calabria
Italy

Signorino Galvagno
University of Messina
Italy

Claudia Espro
University of Messina
Italy

Editorial Office
MDPI
St. Alban-Anlage 66
4052 Basel, Switzerland

This is a reprint of articles from the Special Issue published online in the open access journal *Catalysts* (ISSN 2073-4344) from 2018 to 2019 (available at: https://www.mdpi.com/journal/catalysts/special_issues/catalytic_biomass_derived).

For citation purposes, cite each article independently as indicated on the article page online and as indicated below:

LastName, A.A.; LastName, B.B.; LastName, C.C. Article Title. *Journal Name* **Year**, *Article Number*, Page Range.

ISBN 978-3-03921-914-8 (Pbk)
ISBN 978-3-03921-915-5 (PDF)

Contents

About the Special Issue Editors

Francesco Mauriello (PhD in Industrial Chemistry) is currently an Assistant Professor at the University of Reggio Calabria (Italy). In 2017 he received the National Academic Qualification to serve as Associate Professor of Fundamentals of Chemical Technologies (Academic Discipline: Chim/07 - Academic Recruiting Field: 03/B2). His research activities are focused in the field of synthesis and physico–chemical characterization of heterogeneous catalysts for the reductive transformation of renewable raw materials into added value chemical compounds and bio-fuels via hydrogenolysis reactions. Specifically, he is experienced in the evaluation of catalytic properties (activity, selectivity, and stability) of metallic and bimetallic systems supported on oxide materials. His recent appointments include visiting Assistant Professor at iCAT, formely known as the Catalysis Research Center, Hokkaido University, Japan, from 2012 to 2013 and CAT Catalytic Center, Aachen University, Germany, in 2012. He received several awards/fellowships, including (i) the DAAD Research fellowship (2012), the IUPAC Young Scientist Award (43rd IUPAC World Chemistry Congress, 2011) and (iii) the "Marco Polo fellowship" (Early Stage Researchers mobility, 2007).He is the author of 41 scientific publications in international journals, two book chapters (1 in publication) and his H index is currently 14 (SCOPUS, October 2019), with more than 500 citations.He is an active member of the Italian Chemical Society (Vice President of the regional section). Since 2018 has been a member of the Editorial Board of the journals Catalysts (MDPI) and Current Catalysis (Bentham Science), and serves as a reviewer for ACS, RSC, Elsevier, Wiley, MDPI, Frontiers and research programs from the Italian Ministry of Research (MIUR), the Poland National Science Centre and the Chilean National Science and Technology Commission (CONICYT—Chile).

Signorino Galvagno has been the University of Messina Coordinator of the Doctoral Course in Materials and Constructions Engineering and Chemistry since November 1994. He became a postdoctoral fellow at Michigan University (USA) in 1976 and an Assistant Professor from 1978 to 1981. From 1980 to 1981 was a senior scientist at the Montedipe of Milan. He is the principal investigator of several national and international research projects. His research activity, carried out in both industrial and academic research laboratories, mainly addresses heterogeneous catalysis and, in particular, noble metals supported on high surface area oxides catalytic systems. A significant part of this interest is focused on the development of selective catalysts for low environmental impact processes and new materials and technologies in the energy sector. He is author/co-author of about 300 papers (180 articles in SCOPUS/WOS indexed international journals and 120 communications and conference proceedings at national and international conferences; as well as one patent and an H-Index of 46).

Claudia Espro has been an Assistant Professor at the University of Messina since November 2014. In 2000 she obtained a postgraduate qualification in Chemical Process Technologies. In 2007 she received her PhD from the University of Messina. Claudia Espro conducts research in heterogeneous catalysis and development of novel catalytic green processes. She is pursuing fundamental and applied research objectives. The main topic of her scientific activity lies in the catalytic conversion of natural gas and light alkanes into intermediates, fuels and chemicals of higher added value. Her scientific activity is also devoted to other research subjects related to the conversion of renewable biomass for the production of bulk chemicals. She has participated in various research projects in the framework of national and international research programmes and R&D activities of chemical and petrochemical industries. She is the author/co-author of about 120 papers (55 articles in SCOPUS/WOS indexed international journals and 65 communications and conference proceedings at national and international conferences; as well as one patent and an H-Index of 15).

catalysts

MDPI

Editorial

Catalytic Processes for The Valorization of Biomass Derived Molecules

Claudia Espro [1],* and Francesco Mauriello [2],*

[1]   Dipartimento di Ingegneria, Università di Messina, Contrada di Dio–Vill. S. Agata, I-98166 Messina, Italy
[2]   Dipartimento DICEAM, Università Mediterranea di Reggio Calabria, Loc. Feo di Vito,
      I-89122 Reggio Calabria, Italy
*   Correspondence: espro@unime.it (C.E.); francesco.mauriello@unirc.it (F.M.); Tel.: +39-090-6765264 (C.E.);
      +39-0965-1692278 (F.M.)

Received: 30 July 2019; Accepted: 6 August 2019; Published: 8 August 2019

1. Introduction

Industrial chemistry is changing its fossil distinctiveness into a new green identity by using renewable resources. Biomasses, produced and used in a cyclical way, constitute an important environmentally friendly resource for the production of energy, chemicals, and biofuels [1,2]. To this regard, abundant and inedible lignocellulosic biomasses have attracted a lot of attention being not in competition with agricultural land and food production and, therefore, representing renewable feedstocks for modern biorefineries. The three main components of lignocellulosic biomasses are cellulose, hemicellulose, and lignin, which can be converted into energy (biogas and H_2), liquid biofuels, and into a pool of platform molecules including sugars, polyols, alcohols, aldehydes, ketones, ethers, esters, acids, and aromatics compounds [3–7]. However, in order to develop efficient catalytic processes for the selective production of desired products from lignocellulose, a deep understanding of the molecular aspects of the basic chemistry and reactivity of biomass derived molecules is still necessary.

This Special Issue aims to cover recent progresses in the catalytic valorization of cellulose, hemicellulose and lignin model molecules promoted by novel heterogeneous systems for the production of energy, fuels and chemicals. In this context, it is worth to highlight some recent research advances. Among many, hydrogenation/hydrogenolysis [8,9] and transfer hydrogenolysis [10–12] of cellulose, hemicellulose, lignin, C6-C3 polyols, furan derivatives and aromatic ethers represent a core technology of modern biorefineries. Glycerol (C3 polyol) and glycidol (2,3-epoxy-1-propanol) can be converted into hydrogen (via APR process), C3-C1 alcohols, as well as to cyclic acetals and ketals [13–16]. Accordingly, reforming processes are surely a new way for the production of H_2 and liquid hydrocarbons from lignocellulosic biomasses or their derived molecules [17–20]. At the same time, biomasses can be directly converted into liquid fuels and bio-oils via pyrolysis (thermal degradation process in absence of added oxygen) [21] or used as starting materials for the production of vegetable oils that are efficiently used in several energetic application [22,23]. Finally, a recent trend is surely the production of aromatics, including BTX compounds (benzene-toluene-xylene), starting from lignin, sugars and aromatic ethers, and esters in the framework of the so-called "lignin-first biorefinery" [24–28].

2. The Contents of the Special Issue

In this special issue, several fields of research on the catalytic valorization of biomasses and their relative molecules are covered. Therefore, we would like to sincerely thank all the authors who contributed with their excellent contributions to this special issue that includes six articles (three reviews among them).

Martín et al. propose a natural zeolite (Chilean) as an innovative catalyst for bio-oil upgrade processes [29]. The results clearly show that Chilean-zeolites efficiently increase both quality and stability of the bio-oil obtained from the catalytic pyrolysis of the Chilean native oak. In particular, zeolite acid sites allow the decrease of oil viscosity as a consequence of the increase of the concentration of hydrocarbons, alcohols and aldehydes during the storage practice. At the same time, the presence of Brønsted acid sites on Chilean-zeolites promotes the reduction of carbonyl and alcoholic groups of bio-oil, even after storage.

The hydro-isomerization upgrading of vegetable oil-based insulating oil was presented by Dieu-Phuong Phan and Eun Yeol Lee [30]. In their review, they demonstrated that vegetable oils can be a valid feedstock of insulating oils for electric transformers presenting the effect of (i) metal phase, (ii) acid sites, and (iii) pore structure on the catalytic hydroisomerization processes. At the same time, the influence of blending processes on the physico-chemical properties of these alternative oils are also presented. Moreover, in their contribution, authors pointed out some of the drawbacks related to vegetable oil-based insulating oils (e.g., high pour points, minor aging, and higher viscosity).

Prof. Jose A. Lopez-Sanchez and co-workers demonstrated the influence of alkaline treatment of H-ZSM-5 catalyst for the production of p-xylene and other aromatic compounds starting from bioderived sugars and ethylene [31]. Authors show that the alkaline treatment allows obtaining a series of catalysts characterized by a mesoporous structure preserving the typical crystallinity of the H-ZSM-5 based zeolites. The key factor in driving the production of p-xylene from 2,5-dimethylfuran in high activity/selectivity was found to be the right compromise between acidity and mesoporosity of the alkaline H-ZSM-5 catalyst.

The use of an acid heterogeneous catalyst (Nafion NR50) for the synthesis of solketal from bio-glycidol was presented by Cucciniello et al. [32]. The Nafion NR50 system was found to be very efficient even at a very low catalyst loading, allowing a quantitative acetalization of glycidol into solketal under relative mild reaction conditions. Moreover, the catalyst can be re-used several times without any significant decrease in activity/selectivity.

The research group of Prof. Cavani and co-workers contributed a review on the use of mixed-oxide catalysts for the chemical-loop reforming (CRL) of bioethanol [33]. Authors pointed out how the different preparation methods drive the reactivity of spinels in the CLR of bioethanol. In particular, M-Fe_2O_4 ferrospinels ($M = Cu$, Co, Mn) were found to be very active in H_2 production using several building block chemicals that can be obtained from ethanol (acetone, acetaldehyde, and C4 compounds) in the second loop. Furthermore, an easy recovery and reuse of the initial M-Fe_2O_4 can be done.

In another review, the use of transfer hydrogenolysis approach for the reductive upgrading of lignocellulosic biomasses is presented [12]. The reductive valorization of cellulose, hemicellulose, lignin, and their relative derived/model molecules in absence of added H_2 allows the production of several added value chemicals including acids, ethers, esters, aromatics, polyols, and alcohols. Furthermore, the use of simple organic molecules such as methanol, ethanol, 2-propanol, and formic acid as an indirect H-source may solve most of the problems related to the classic use of high-pressure molecular hydrogen (including safety hazards, expensive infrastructures, and costs related to the transport/storage of pressurized H_2).

Funding: This research received no external funding.

Conflicts of Interest: All authors declare no conflict of interest.

References

1. Besson, M.; Gallezot, P.; Pinel, C. Conversion of Biomass into Chemicals over Metal Catalysts. *Chem. Rev.* **2014**, *114*, 1827–1870. [CrossRef] [PubMed]
2. Zhou, C.-H.; Xia, X.; Lin, C.-X.; Tong, D.-S.; Beltramini, J. Catalytic conversion of lignocellulosic biomass to fine chemicals and fuels. *Chem. Soc. Rev.* **2011**, *40*, 5588–5617. [CrossRef] [PubMed]
3. Espro, C.; Gumina, B.; Paone, E.; Mauriello, F. Upgrading Lignocellulosic Biomasses: Hydrogenolysis of Platform Derived Molecules Promoted by Heterogeneous Pd-Fe Catalysts. *Catalysts* **2017**, *7*, 78. [CrossRef]

4. Chen, X.; Guan, W.; Tsang, C.-W.; Hu, H.; Liang, C. Lignin Valorizations with Ni Catalysts for Renewable Chemicals and Fuels Productions. *Catalysts* **2019**, *9*, 488. [CrossRef]

5. Shrotri, A.; Kobayashi, H.; Fukuoka, A. Cellulose Depolymerization over Heterogeneous Catalysts. *Acc. Chem. Res.* **2018**, *51*, 761–768. [CrossRef] [PubMed]

6. Delidovich, I.; Leonhard, K.; Palkovits, R. Cellulose and hemicellulose valorisation: An integrated challenge of catalysis and reaction engineering. *Energy Environ. Sci.* **2014**, *7*, 2803–2830. [CrossRef]

7. Xu, C.; Arancon, R.A.D.; Labidi, J.; Luque, R. Lignin depolymerisation strategies: Towards valuable chemicals and fuels. *Chem. Soc. Rev.* **2014**, *43*, 7485–7500. [CrossRef]

8. Ruppert, A.M.; Weinberg, K.; Palkovits, R. Hydrogenolysis goes bio: From carbohydrates and sugar alcohols to platform chemicals. *Angew. Chem. Int. Ed.* **2012**, *51*, 2564–2601. [CrossRef]

9. Li, N.; Wang, W.; Zheng, M.; Zhang, T. General Reaction Mechanisms in Hydrogenation and Hydrogenolysis for Biorefining. In *Catalytic Hydrogenation for Biomass Valorization*; Rinaldi, R., Ed.; Royal Society of Chemistry: Cambridge, UK, 2015; pp. 22–50.

10. Wang, D.; Astruc, D. The Golden Age of Transfer Hydrogenation. *Chem. Rev.* **2015**, *115*, 6621–6686. [CrossRef]

11. Gilkey, M.J.; Xu, B. Heterogeneous Catalytic Transfer Hydrogenation as an Effective Pathway in Biomass Upgrading. *ACS Catal.* **2016**, *6*, 1420–1436. [CrossRef]

12. Espro, C.; Gumina, B.; Szumelda, T.; Paone, E.; Mauriello, F. Catalytic Transfer Hydrogenolysis as an Effective Tool for the Reductive Upgrading of Cellulose, Hemicellulose, Lignin, and Their Derived Molecules. *Catalysts* **2018**, *8*, 313. [CrossRef]

13. Zhou, C.H.; Zhao, H.; Tong, D.S.; Wu, L.M.; Yu, W.H. Recent advances in catalytic conversion of glycerol. *Catal. Rev.* **2013**, *55*, 369–453. [CrossRef]

14. Mauriello, F.; Vinci, A.; Espro, C.; Gumina, B.; Musolino, M.G.; Pietropaolo, R. Hydrogenolysis vs. aqueous phase reforming (APR) of glycerol promoted by a heterogeneous Pd/Fe catalyst. *Catal. Sci. Technol.* **2015**, *5*, 4466–4473. [CrossRef]

15. Vicente, G.; Melero, J.A.; Morales, G.; Paniagua, M.; Martín, E. Acetalisation of bio-glycerol with acetone to produce solketal over sulfonic mesostructured silicas. *Green Chem.* **2010**, *12*, 899–907. [CrossRef]

16. Samoilov, V.O.; Onishchenko, M.O.; Ramazanov, D.N.; Maximov, A.L. Glycerol Isopropyl Ethers: Direct Synthesis from Alcohols and Synthesis by the Reduction of Solketal. *ChemCatChem* **2017**, *9*, 2839–2849. [CrossRef]

17. Cortright, R.D.; Davda, R.R.; Dumesic, J.A. Hydrogen from catalytic reforming of biomass-derived hydrocarbons in liquid water. *Nature* **2002**, *418*, 964–967. [CrossRef] [PubMed]

18. Davda, R.R.; Dumesic, J.A. Catalytic reforming of oxygenated hydrocarbons for hydrogen with low levels of carbon monoxide. *Angew. Chem. Int. Ed.* **2003**, *42*, 4068–4071. [CrossRef]

19. Vozniuk, O.; Agnoli, S.; Artiglia, L.; Vassoi, A.; Tanchoux, N.; Di Renzo, F.; Granozzi, G.; Cavani, F. Towards an improved process for hydrogen production: The chemical-loop reforming of ethanol. *Green Chem.* **2016**, *18*, 1038–1050. [CrossRef]

20. Hočevar, B.; Grilc, M.; Likozar, B. Aqueous Dehydration, Hydrogenation, and Hydrodeoxygenation Reactions of Bio-Based Mucic Acid over Ni, NiMo, Pt, Rh, and Ru on Neutral or Acidic Catalyst Supports. *Catalysts* **2019**, *9*, 286. [CrossRef]

21. Demirbas, A.; Arin, G. An overview of biomass pyrolysis. *Energy Source* **2002**, *24*, 471–482. [CrossRef]

22. Nagashree, A.N.; Champa, V.; Sumangala, B.V.; Nagabhushana, G.R. Suitability of natural vegetable seed oil as liquid dielectric coolant in an insulation system. In Proceedings of the 2015 International Conference on Emerging Research in Electronics, Computer Science and Technology (ICERECT), Mandya, India, 17–19 December 2015; pp. 429–434.

23. McShane, C.P. Relative properties of the new combustion-resist vegetable-oil-based dielectric coolants for distribution and power transformers. *IEEE Trans. Ind. Appl.* **2001**, *37*, 1132–1139. [CrossRef]

24. Mauriello, F.; Paone, E.; Pietropaolo, R.; Balu, A.M.; Luque, R. Catalytic transfer hydrogenolysis of lignin-derived aromatic ethers promoted by bimetallic Pd/Ni systems. *ACS Sustain. Chem. Eng.* **2018**, *6*, 9269–9276. [CrossRef]

25. Cozzula, D.; Vinci, A.; Mauriello, F.; Pietropaolo, R.; Müller, T.E. Directing the Cleavage of Ester C–O Bonds by Controlling the Hydrogen Availability on the Surface of Coprecipitated Pd/Fe$_3$O$_4$. *ChemCatChem* **2016**, *8*, 1515–1522. [CrossRef]

26. Paone, E.; Espro, C.; Pietropaolo, R.; Mauriello, F. Selective arene production from transfer hydrogenolysis of benzyl phenyl ether promoted by a co-precipitated Pd/Fe$_3$O$_4$ catalyst. *Catal. Sci. Technol.* **2016**, *6*, 7937–7941. [CrossRef]

27. Cao, Z.; Dierks, M.; Clough, M.T.; de Castro, I.B.D.; Rinaldi, R. A convergent approach for a deep converting lignin-first biorefinery rendering high-energy-density drop-in fuels. *Joule* **2018**, *2*, 1118–1133. [CrossRef] [PubMed]

28. Renders, T.; den Bossche, G.V.; Vangeel, T.; Van Aelst, K.; Sels, B. Reductive catalytic fractionation: State of the art of the lignin-first biorefinery. *Curr. Opin. Biotechnol.* **2019**, *56*, 193–201. [CrossRef] [PubMed]

29. Alejandro Martín, S.; Cerda-Barrera, C.; Montecinos, A. Catalytic Pyrolysis of Chilean Oak: Influence of Brønsted Acid Sites of Chilean Natural Zeolite. *Catalysts* **2017**, *7*, 356. [CrossRef]

30. Phan, D.-P.; Lee, E.Y. Catalytic Hydroisomerization Upgrading of Vegetable Oil-Based Insulating Oil. *Catalysts* **2018**, *8*, 131. [CrossRef]

31. McGlone, J.; Priecel, P.; Da Già, L.; Majdal, L.; Lopez-Sanchez, J.A. Desilicated ZSM-5 Zeolites for the Production of Renewable *p*-Xylene via Diels–Alder Cycloaddition of Dimethylfuran and Ethylene. *Catalysts* **2018**, *8*, 253. [CrossRef]

32. Ricciardi, M.; Falivene, L.; Tabanelli, T.; Proto, A.; Cucciniello, R.; Cavani, F. Bio-Glycidol Conversion to Solketal over Acid Heterogeneous Catalysts: Synthesis and Theoretical Approach. *Catalysts* **2018**, *8*, 391. [CrossRef]

33. Vozniuk, O.; Tabanelli, T.; Tanchoux, N.; Millet, J.-M.M.; Albonetti, S.; Di Renzo, F.; Cavani, F. Mixed-Oxide Catalysts with Spinel Structure for the Valorization of Biomass: The Chemical-Loop Reforming of Bioethanol. *Catalysts* **2018**, *8*, 332. [CrossRef]

catalysts

MDPI

Article

Catalytic Pyrolysis of Chilean Oak: Influence of Brønsted Acid Sites of Chilean Natural Zeolite

Serguei Alejandro Martín [1,2,*], Cristian Cerda-Barrera [3] and Adan Montecinos [3]

[1]   Wood Engineering Department, Faculty of Engineering, University of Bío-Bío (UBB), 4030000 Concepción, Chile
[2]   Nanomaterials and Catalysts for Sustainable Processes (NanoCat$_p$PS), UBB, 4030000 Concepción, Chile
[3]   School of Industrial Processes Engineering, Temuco Catholic University, 4780000 Temuco, Chile;
      ccerda@uct.cl (C.C.-B.); amontecinos2010@alu.uct.cl (A.M.)
*   Correspondence: salejandro@ubiobio.cl; Tel.: +56-41-311-1168

Received: 2 November 2017; Accepted: 21 November 2017; Published: 24 November 2017

Abstract: This paper proposes the Chilean natural zeolite as catalyst on bio-oil upgrade processes. The aim of this study was to analyze chemical composition of bio-oil samples obtained from catalytic pyrolysis of Chilean native oak in order to increase bio-oil stability during storage. In order to identify chemical compounds before and after storage, biomass pyrolysis was carried out in a fixed bed reactor at 623 K and bio-oil samples were characterized by gas chromatography/mass spectrophotometry (GC/MS). A bio-oil fractionation method was successfully applied here. Results indicate that bio-oil viscosity decreases due to active sites on the zeolite framework. Active acids sites were associated with an increment of alcohols, aldehydes, and hydrocarbon content during storage. Higher composition on aldehydes and alcohols after storage could be attributed to the occurrence of carbonyl reduction reactions that promotes them. These reactions are influenced by zeolite surface characteristics and could be achieved via the direct contribution of Brønsted acid sites to Chilean natural zeolite.

Keywords: bio-oil upgrade; Brønsted acids sites; Chilean natural zeolites; GC/MS characterization

1. Introduction

The increase in energy consumption as a consequence of global economic expansion, renewable energies must be widely explored in order to assure sustainable development of human race. In addition, the reduction of fossil fuel resources, the rise of atmospheric carbon dioxide levels, and the gradually emerging consciousness about environmental degradation has promoted novel strategies for the sustainable production of fuels, mainly from renewable sources. Among them, biomass possess an adequate availability for the large-scale sustainable production of liquid fuels [1,2], contributes to about 12% of the world's primary energy supply, and is claimed as the only renewable substitute of organic petroleum [3].

Biomass pyrolysis (thermal degradation of biomass by heat in the absence of oxygen), which results in charcoal (solid), bio-oil (condensed liquid), and fuel gas products, offers a convenient way to obtain liquid fuels, and it has become an essential area for the research and development of new processes [3]. Although bio-oil is a potential fuel, it cannot be used as such without prior upgrading due to its high viscosity, low heating value, corrosiveness, instability, and water content [4]. Among the different alternative methods that improve the bio-oil quality by lowering its oxygen content, zeolite cracking is one of the most promising options [5].

The preferred catalysts for bio-oil upgrading processes are synthetic zeolites, such as ZSM-5, HZSM-5, zeolite A, and zeolite Y [6]. It has been reported before that acidic zeolites (H-Y and H-ZSM5) increased the desirable chemical compounds in bio-oil, such as phenols, furans, and hydrocarbons, and reduced the undesired compounds, such as acids [7]. However, their high cost constitutes a disadvantage. That is why some researchers have investigated the use of natural zeolites on bio-oil upgrading processes [8–10].

Ates et al. (2005) confirmed that pyrolytic liquid yield increases when natural zeolite (clinoptilolite) is used as a catalyst, in comparison to non-catalyst tests. Messina et al. (2017) investigated the in-situ catalytic pyrolysis of peanut shells using natural (clinoptilolite) and modified zeolites to study the deoxygenation degree on bio-oil samples. Modified zeolite samples were obtained by ionic exchange with NH_4Cl in order to develop Brønsted acid sites on framework [5]. Rajić et al. (2013) studied the catalytic activity of Na-rich and MO-containing natural clinoptilolite (MO = nanoparticles of NiO, ZnO, or Cu_2O) in the hardwood lignin pyrolysis [11]. As seen, several studies have been performed to identify interesting biomass pyrolysis derivatives that could be used as bio-oils directly or as raw materials for chemical industries [12,13]. However, there have been no previous reports related to native oak or natural zeolites from Chile. Therefore, the main goal of this article is to analyze the chemical composition of bio-oil obtained from the pyrolysis of Chilean native oak so as to study the bio-oil stability during storage.

2. Results and Discussion

2.1. Biomass and Natural Zeolite Characterization

Biomass is a composite material of oxygen-containing organic polymers (cellulose, hemicelluloses, lignins, and extractives) and inorganic minerals. The weight percent of these components varies from one species to other. Chilean oak was characterized according to the aforementioned procedures, and the results of proximate (dry basis), ultimate, and elemental Analyses are summarized in Table 1.

Table 1. Biomass proximate (dry basis), ultimate, and elemental analyses.

Volatile Matter [wt %]	Fixed Carbon [wt %] [a]	Ash [wt %]	GCV [MJ kg^{-1}] [a]	GCV [MJ kg^{-1}] [b]	Cellu Lose [%]	Extrac Tives [%]	Lignin [%]	Hemi Cellulose [%] [c]	C	H	O	N
85.74	12.62	1.64	20.72	24.93	35.38	1.97	27.10	35.55	47.3	6.36	46.34	-

[a] Dried at 378 K by 12 h; [b] Bio-char obtained by pyrolysis at 623 K by 30 min. Gross calorific value (GCV); [c] by difference.

Results obtained here are quantitatively similar to those reported previously, taking into consideration that the main structural chemical components (carbohydrate polymers and oligomers) constitute about 65–75%, lignin 18–35%, and organic extractives and inorganic minerals usually 4–10% [14].

Natural zeolite was characterized via different techniques, as mentioned. Characterization results are summarized in Table 2.

Table 2. Physico-chemical characterization of Chilean natural zeolite (NZ).

S [m^2 g^{-1}]	SiO$_2$ [a]	Al$_2$O$_3$ [a]	Na$_2$O [a]	CaO [a]	K$_2$O [a]	MgO [a]	TiO$_2$ [a]	Fe$_2$O$_3$ [a]	MnO [a]	P$_2$O$_5$ [a]	CuO [a]	Si/Al
168.17	71.61	15.18	2.0	3.43	2.03	0.74	0.61	3.99	0.06	0.12	0.03	4.72

[a] by XRF (X-ray fluorescence) (% w/w).

Zeolite surface area (S) was calculated with the Langmuir adsorption model using nitrogen adsorption data. The sample was outgassed at 623 K for 12 h prior to the measurements. XRD (X-ray powder diffraction) patterns indicated that the zeolite sample was highly crystalline, showing characteristic peaks of mordenite and quartz. The mordenite framework consists of two channel systems: the perpendicular channel system has 12 MR pores with dimensions of 6.7×7.0 Å, and the parallel channel system has 8 MR pores with dimensions of 2.9×5.7 Å. These channels are interconnected with small side pockets with a diameter of 2.9 Å [10]. X-ray fluorescence (XRF) identifies compensating cations (Na, K, Mg, Ca, Mg, Fe, and P) in the zeolite framework. The low Si/Al ratio in the zeolite framework was similar to those of Chilean natural zeolites reported previously [15].

2.2. Pyridine Adsorption Followed by DRIFT

The DRIFT (Diffuse Reflectance Infra-red Fourier Transform) technique allowed for the identification of Brønsted and Lewis acid sites on natural zeolite, as shown in Figure 1. A progressive thermal desorption procedure was conducted in a DRIFT chamber in order to evaluate Brønsted and Lewis acid site strength. Pyridine-saturated samples were subjected to vacuum and temperatures of 373 K, 473 K, 573 K, and 673 K, starting from room temperature (RT). After each heating stage, samples were cooled down to 293 K and respectively spectrum was recorded. Interaction of the pyridine molecule with the Brønsted (Py-B) and Lewis (Py-L) acid sites were confirmed by the registered peaks, near 1540 and 1450 cm^{-1}, respectively. The band at 1456 cm^{-1} can be assigned to the adsorbed pyridine at the Lewis acid sites. In the same way, the band at 1539 cm^{-1} is due to the Brønsted acid sites, whereas the band at 1488 cm^{-1} could be attributed to both Lewis and Brønsted acidity. On the basis of the above, the presence of at least one type of Lewis acid site and one type of Brønsted site was concluded, as has been reported previously [16–18].

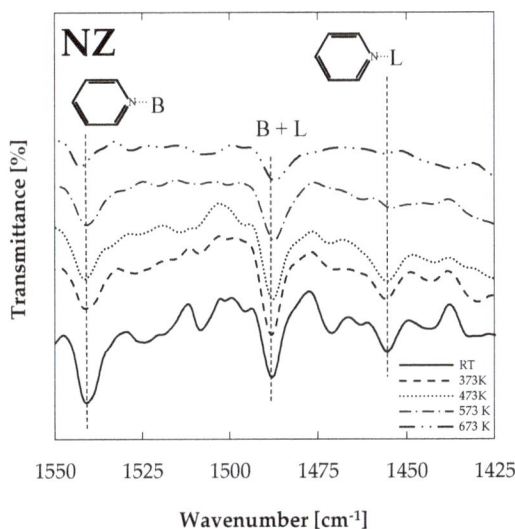

Figure 1. DRIFT spectrum of adsorbed pyridine on natural zeolite.

2.3. Thermal Behavior of Biomass

Figure 2 shows TG/DTG/DTA (Thermogravimetric/Derivative Thermogravimetric/Differential Thermogravimetric Analysis) profiles of thermogravimetric analysis of biomass (oak) in a nitrogen atmosphere. Three zones were identified: drying (T < 383 K) [not shown here]; pyrolysis (476–668 K), and passive pyrolysis (T > 668 K).

As regards wood composition, DTG shows two peaks corresponding to hemicellulose and cellulose in the active pyrolysis zone and the tailing zone associated with lignin degradation in the passive pyrolysis zone, as reported elsewhere [19].

The DTG profile calculated for biomass pyrolysis showed a shoulder, a peak, and a tail, clearly indicating that multiple reactions were involved. The overlapping peaks correspond to hemicellulose and cellulose degradation in the active pyrolysis stage, and the tailing zone is associated with lignin decomposition in the passive pyrolysis stage. Maximum degradation of hemicellulose occurs at $T_1 = 576$ K, which was calculated from the DTG shoulder and the D^{2}TG (second derivative of the weight loss) valley represented in Figure 2. On the other hand, $T_m = 628$ K indicates the temperature of the cellulose maximum degradation rate (0.01 mg·s^{-1}). A 59% of initial mass was removed at this temperature.

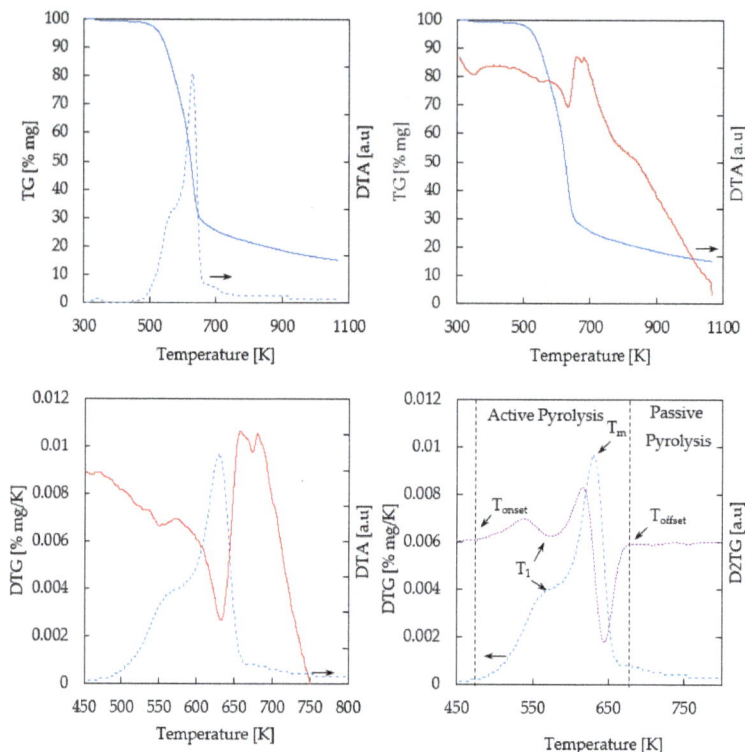

Figure 2. Thermogravimetric profiles of biomass sample. Solid blue line: TG; Solid red line: DTA; Dotted blue line: 1st TG derivative; Dotted purple line: 2nd TG derivative.

The second derivative of the weight loss (D^{2}TG) was applied in order to estimate temperature ranges for thermal degradation of Chilean oak main components (hemicellulose, cellulose, and lignin). Close to zero values on the D^{2}TG curve were established as the starting (T_{onset}) and end (T_{offset}) temperatures of the involved steps. Then, the active pyrolysis occurred between 476 and 668 K, releasing 84% (α_{offset}) of the volatile material from the oak samples. Considering this, all pyrolysis experiments were conducted at 623 K. Finally, after the TGA temperature program was carried out, the residual carbonaceous material was 13.9%.

2.4. Product Distribution during Chilean Oak Pyrolysis

Product distribution (bio-oil, permanent gases, and bio-char) for studied biomass (oak) were calculated from mass balance. Results are summarized in Table 3.

Table 3. Products distribution for oak pyrolysis at 623 K.

	Product Yield [% w]		
	Bio-Oil	**Bio-Char**	**Permanent Gases**
Non-Catalytic	37.58	31.44	30.98
Catalytic	44.18	29.78	26.03

Exp. Conditions: m_{oak} = 9 gr. N$_2$ flow = 60 cm$^3 \cdot$min^{-1}. Ramp rate = 10 K$\cdot$min^{-1}.

Catalytic experiments registered a higher bio-oil yield (44.1%) than non-catalytic assays (37.5%), in accordance with a previous study [8]. This also indicates a difference in chemical composition of the gaseous stream obtained from catalytic runs, as a consequence of zeolite acid site interaction with derived chemical compounds.

2.5. Viscosity Variation during Bio-Oil Storage at Different Temperatures

Bio-oil samples obtained from non-catalytic and catalytic pyrolysis were stored (in sealed amber vials) at different temperatures (277 K and 353 K) for 507 h (3 weeks), in order to evaluate the effect of temperature on viscosity of samples. Results shown in Figure 3 represent the difference ($\Delta\mu$) between bio-oil viscosity at the end (μ_{507}) and after 3 h (μ_3) of storage.

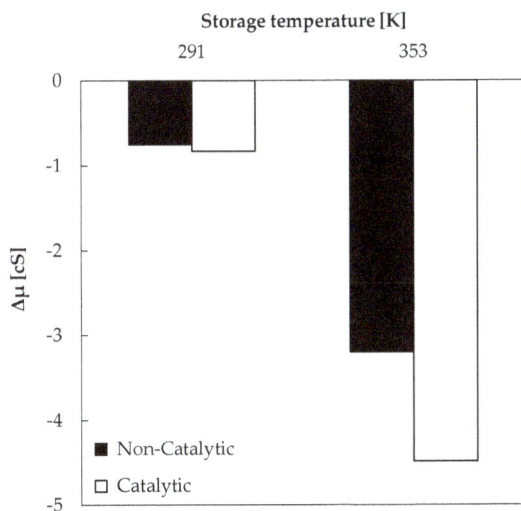

Figure 3. Viscosity decrease during bio-oil storage (507 h).

A general trend was observed here: the viscosity variation became higher when storage temperature increased. A greater viscosity decrease was registered for catalytic bio-oil stored at 353 K. Taking into consideration that the highest initial viscosity value (11.8 cS) was registered for catalytic bio-oil stored at 353 K, the viscosity variation could be associated with changes in bio-oil chemical composition during storage. On the other hand, bio-oil obtained from non-catalytic tests and stored at 353 K registered a viscosity of 8.78 cS after 3 h of storage. Viscosity decreases with high levels of water and low-molecular-weight compounds, and increases with high lignin concentrations of pyrolysis materials and insoluble solids. Below 353 K, viscosity increases linearly with temperature, in accordance with Figure 3 and the behavior of Newtonian fluids. However, over 353 K, some bio-oil compounds are volatilized and polymerization may occur [20,21]. Therefore, storage tests in this study were conducted with 353 K as the maximum temperature.

2.6. Chemical Composition of Stored Bio-Oil Samples

Several compounds were identified on bio-oil samples from oak pyrolysis at 623 K. In order to clarify the distribution of compounds among different fractions, compounds were distributed in different families: acids, alcohols, aldehydes, ketones, esters, ethers, and hydrocarbons, as commonly reported in other studies [22–26]. Table 4 shows the qualitative composition of bio-oil samples obtained from (non-catalytic and catalytic) pyrolysis and stored at 277 K for 3 months.

Table 4. Qualitative analysis of bio-oil obtained from (non-catalytic and catalytic) oak pyrolysis.

	Oak Pyrolysis	
	Non-Catalytic	Catalytic
Compounds registered on chromatogram	90	117
Oxygenated compounds	43	54
Non-oxygenated compounds	15	14
Unknown	32	49

A higher number of chemical compounds were registered on chromatograms of fractionated bio-oil samples from catalytic assays. Those compounds might be associated with additional surface reactions promoted by active acid sites in the natural zeolite framework. Acid sites thus might play a key role in catalytic pyrolysis, leading to the transformation of original species obtained from non-catalytic oak pyrolysis. Furthermore, natural zeolites with lower Si/Al ratios have been shown to be effective in cracking reactions during pyrolysis reactions [5,27].

Figure 1 shows collected spectra after each heating procedure applied to pyridine-saturated samples. A higher peak was registered at 1539 cm^{-1}, and a very small peak was observed at 1456 cm^{-1} for outgassed samples at 673 K, confirming the higher strength of Brønsted acid sites in comparison with Lewis acid sites in the natural zeolite sample. Thus, stronger Brønsted acid sites will cause bond breakage, transforming oxygenated compounds into other species. Stronger sites are needed here, considering that oxygenated compound removal is limited by the C–O bond breakage possibility and also the C–O bond strength varies from one family to other (e.g., the strength of C–O bonds attached to an aromatic ring in phenols or to aryl ethers is greater than that of bonds attached to an aliphatic C in alcohols or aliphatic ethers [28]). Brønsted acid sites are claimed as responsible for acid catalyzed reactions such as cracking, dimerization, cyclization, and dehydrocyclization [29].

The high oxygen content (about 35–40% and distributed in several compounds) in bio-oils is responsible for the low calorific value, corrosiveness, and instability [4,30]. On the other hand, a higher water content, a less water-insoluble components, and an increase in alcohols composition reduces bio-oil viscosities during storage [31,32]. Boucher et al. (2000) demonstrated that the addition of methanol reduces the density and viscosity of bio-oils, increasing the stability [33]. It has been reported that methanol or ethanol additions to bio-oil samples constitute a simple method of viscosity control [34].

A semi-quantitative GC/MS analyses of fractionated bio-oil samples was conducted here in order to identify chemical compounds. Figure 4 shows the area percentage of the main compound families in bio-oil samples, obtained from non-catalytic and catalytic pyrolysis runs. It can be seen that some compounds families (alcohols, aldehydes, and hydrocarbons) are higher on catalytic runs. On the other hand, ketone, ester, and ether concentrations decrease from non-catalytic to catalytic runs.

After three months of storage, the area percentages of the main compound families from the non-catalytic bio-oil sample were as follows: alcohols (7.28%), aldehydes (2.59%), ketones (19.84%), esters (14.82%), hydrocarbons (42.59%), and ethers (12.84%). On the other hand, the catalytic bio-oil sample composition was as follows: alcohols (31.84%), aldehydes (7.61%), ketones (3.86%), esters (9.56%), hydrocarbons (46.02%), and ethers (1.11%). As Figure 4 shows, the composition of hydrocarbons, aldehydes, and alcohols of catalytic bio-oil samples are higher than those obtained for non-catalytic samples. The higher viscosity variation registered (see Figure 3) for catalytic bio-samples during storage could be associated with a higher alcohol content, considering the aforementioned contribution of alcohols to bio-oil viscosity. Furthermore, the higher composition of hydrocarbons could be achieved via the direct contribution of acid sites to the zeolite samples. Ketone composition was lower in catalytic bio-oil samples. Other authors have previously reported that biomass pyrolysis using acidic zeolites as a catalyst produces fewer ketones than non-catalytic runs [29]. It is suggested in the literature that compositional changes in bio-oil samples are due to potential aging reactions (e.g., organic acids form esters with alcohols removing water; aldehydes form oligomers and resins, aldehydes and phenolic compounds form resins and water, aldehydes and proteins form oligomers) [35].

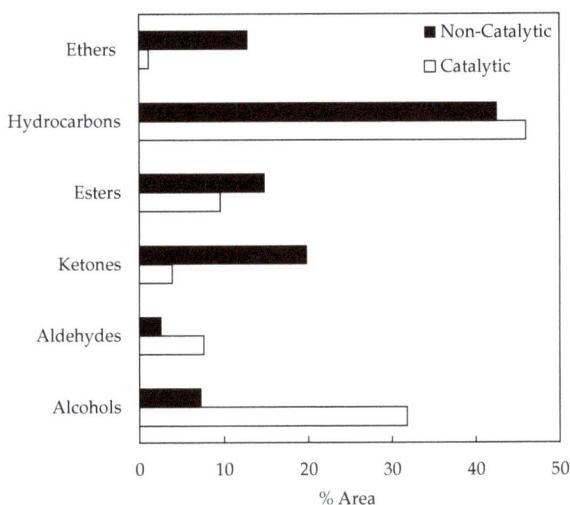

Figure 4. Compound families in stored bio-oil samples.

A detailed register of main compounds on different families is shown on Table 5. Identified compounds in bio-oils are distributed in more than 200 species, depending on the raw material and pyrolysis conditions (temperature, residence time and heating rate). Compounds summarized here have also been reported in other articles [20,22,23].

Listed compounds in Table 5 are mainly oxygenated chemicals with different functional groups including hydroxyl, phenolic hydroxyl, carboxyl, carbonyl, methoxy, ethoxy, oxygen-containing heterocyclic, and unsaturated double bonds. Those species contain a large portion of unsaturated bonds, promoting additional reactions during storage [36]. In order to study this singularities, the composition of catalytic bio-oil samples were analyzed by the aforementioned GC/MS procedure, before and after three months of storage.

Table 5. Main components of bio-oil obtained from (non-catalytic and catalytic) oak pyrolysis.

* Fam.	Compounds	% Area		Compounds	% Area	
		N–C	Cat.		N–C	Cat.
Alc	1,2-Benzenediol		1.43	(S)-3-Ethyl-4-methylpentanol		22.54
	1,2-Benzenediol, 3-methoxy-	1.25		3-Furanmethanol		1.45
	Cerulignol	12.75	0.57	p-Ethyl guaiacol		19.07
	Guaiacol	3.02	10.92	1-Heneicosanol		34.96
	4-Ethylguaiacol	3.01		2-Methoxy-4-vinylphenol		2.44
	4-Vinylguaiacol	2.05		5-tert-Butylpyrogallol		23.11
	Homoguaiacol	3.82		Furyl alcohol	1.73	4.66
	Isocreosol	0.27	13.99	o-Guaiacol		2.7
	Methoxyeugenol	5.31	46.09	p-Methylguaiacol		3.12
	o-Benzenediol	1.14		Pyrogallol 1-methyl ether		1.18
	Syringol	4.02	29.94			
Ald	2-Heptadecenal		35.96	Butanedial (Succinaldehyde)	0.3	3.61
	Furfural	8.76	9.4	Syringaldehyde	1.53	0.74
	Vanillin	1.38	0.35	Furfural, 5-methyl	1.61	1.75
K	1,2-Cyclopentanedione, 3-methyl	1.15	10.42	2,4,6-Tris(1,1-dimethylethyl)-4-methylcyclohexa-2,5-dien-1-one	5.34	
	2-Cyclopenten-1-one, 2-hydroxy	1.05	6.23	Acetosyringone		1.65
	Ethanone, 1-(3-hydroxy-4-methoxyphenyl)	16.1		2-Hydroxy-2-methyl-4-pentanone/Tyranton	82.39	4.34
	4-Hydroxy-2-pentanone	2.72		4-Heptanone, 3-methyl-		0.76
	1,2-Cyclopentanedione		1.34			
Es	Acetol acetate	0.32	0.42	Isopentyl 2-methylpropanoate		61.68
	Butanoic acid, 3-methylbutyl ester	21.15		Isopropyl acetate		3.89
	Propanoic acid, 2-methyl-, 2-methylbutyl ester	59.91				
Et	3,4,5-Trimethoxytoluene	0.2		Furan, tetrahydro-2,5-dimethoxy-	0.1	
	Benzene, 1,4-dimethoxy	0.8		Propane, 2-ethoxy-	69.75	
	Benzene, 1,2,3-trimethoxy-5-methyl-		7.73			

Table 5. *Cont.*

* Fam.	Compounds	% Area		Compounds	% Area	
		N–C	Cat.		N–C	Cat.
	Docosane	28.25	29.22	Pentane, 2-methyl-	3.02	
	Heneicosane	4.11	3.18	Tetracosane	45.17	42.68
	Heptacosane	20.58	19.93	Tricosane	32.51	28.87
	Hexacosane	26.71	30.26	Undecane	2.56	2.59
HC	Hexadecane	4.9		Triacontane	2.71	
	Icosane	13.85	10.55	Butane, 2,2-dimethyl-		79.01
	Nonacosane	9.39	10.21	Cetane		4.91
	Octacosane	13.07	14.52	2-Hexene, 3,5,5-trimethyl		15.78
	Pentacosane	26.18	29.32	Hexane, 2-methyl-	2.02	

* Alc: Alcohols; Ald: Aldehydes; Ke: Ketones; Es: Esters; Et: Ethers; HC: Hydrocarbons; N–C: Non-Catalytic; Cat.: Catalytic.

The data in Figure 5 represent compositional changes of compound families during storage of catalytic bio-oil samples. After the catalytic run (t = 0), the area percentages of compound families of the bio-oil sample were as follows: alcohols (30.14%), aldehydes (2.69%), ketones (35.74%), esters (0.66%), hydrocarbons (16.73%), and ethers (3.48%). On the other hand, the catalytic bio-oil sample (t = 3 months) composition was as follows: alcohols (31.84%), aldehydes (7.61%), ketones (3.86%), esters (9.56%), hydrocarbons (46.02%), and ethers (1.11%). The higher composition of aldehydes and alcohols after storage could be attributed to the occurrence of carbonyl reduction reactions that promotes them. In the same way, ester composition increased during storage probably due to alcohol and acid reactions. Registered acid composition was very low, apparently as a consequence of simultaneous alcohol–acid and ketone–acid reactions, considering that ketone composition also decreased during storage.

Based on theoretical backgrounds, the main reactions responsible for bio-oil aging are as follows: esterification (alcohols with organic acids forming esters and water); transesterification (exchanging of alcohol and acid groups in a mixture of two or more esters); homopolymerization (aldehydes reacting with each other to form polyacetal oligomers and polymers); hydration (aldehydes or ketones mixed with water react to form hydrates); hemiacetal formation (alcohol with an aldehyde forming hemiacetal); acetalization (aldehydes and alcohols reacting to form acetals); and phenol/aldehyde reactions [20,36]. These types of reactions increase the average molecular weight, viscosity, and water content. Moreover, phase separation can eventually occur in the long-term storage of bio-oils [28].

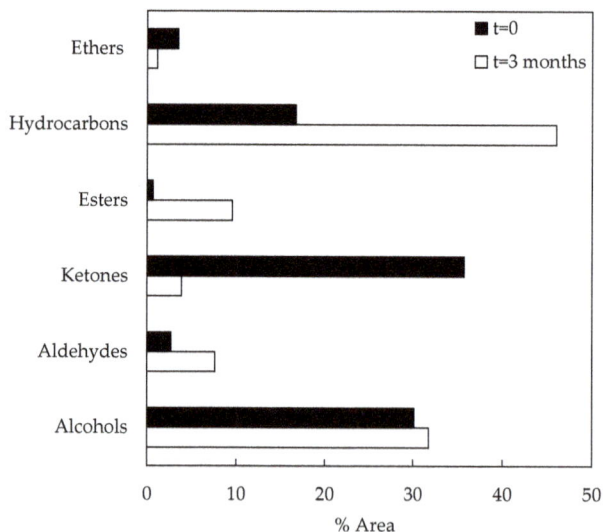

Figure 5. Compound families of catalytic bio-oil samples (before and after storage).

3. Materials and Methods

Biomass samples from Chilean native oak (*Nothofagus obliqua*) were obtained from a single tree, donated by Miraflores Angol Ltda. from Angol, Chile. Samples were subjected to size reduction by sawing, chipping, grinding, and sieving to achieve a granulometry lower than 2 mm. Then, samples were dried at 313 K for 48 h and stored until further use. Biomass properties were obtained by the following standard methods: humidity (UNE-EN 14774); ash content (UNE-EN 14775) and calorific value (UNE-EN 14918). Elemental and proximate analysis were conducted in accordance with UNE-EN 15104 and ASTM D 3172-73(84) standard methods, respectively [37,38]. Thermogravimetric analysis of biomass was performed with a thermal analyzer (Shimadzu DTG-60H, CROMTEK, Santiago, Chile).

Chilean natural zeolite (supplied by Minera FORMAS, Santiago, Chile) was ground and sieved to 0.3–0.425 mm, then rinsed with ultrapure water, oven-dried at 398 K for 24 h, and finally stored in a desiccator until further use. Zeolite samples were characterized via nitrogen absorption at 77 K, X-ray powder diffraction (XRD), and X-ray fluorescence (XRF). Nitrogen adsorption isotherms at 77 K were obtained on a Micromeritics Gemini 3175 (UDEC, Concepción, Chile). XRD was performed with a Bruker AXS Model D4 ENDEAVOR diffractometer (UDEC, Concepción, Chile), equipped with a cupper X-ray tube and an Ni filter, in order to evaluate mineralogical and structural frameworks. XRF, using a RIGAKU Model 3072 spectrometer (UDEC, Concepción, Chile), allowed for the determination of bulk chemical composition of natural and modified zeolites. Characterization assays were conducted according to procedures previously described [15,39,40].

Additionally, zeolite samples were characterized by pyridine adsorption followed by DRIFT (Diffuse Reflectance Infra-red Fourier Transform) in order to confirm acid sites in the zeolite framework. Samples were saturated ex-situ by exposure to pyridine vapors in a closed flask for 12 h. Spectra were obtained using a Nicolet Avatar 370 MCT with a smart collector accessory, a mid/near infrared source, and a mercury cadmium telluride (MCT-A) photon detector at 77 K (liquid N_2). Zeolite samples were mixed with KBr powder (10% w/w zeolite/KBr) prior to pelletizing procedure. Excess of physisorbed pyridine was removed in a vacuum oven prior to sample loading. Samples were placed in the DRIFT cell holder and degasified under vacuum at 373 K for 10 min, in order to eliminate physisorbed moisture during manipulation of samples.

To adsorb water or any other compound, natural zeolite samples were outgassed before the experiments by heating at 823 K for 2 h. Pyrolysis experiments were conducted in a cylindrical reactor using 9 mg of biomass. The heat ramp rate and the N_2 (Indura, 99.998%, Temuco, Chile) flow were 10 K·min^{-1} and 60 cm^3·min^{-1}, respectively. Pyrolysis experiments were carried out using the experimental system shown in Figure 6. Biomass was heated from room temperature to 623 K and was kept isothermal for 30 min. Zeolites were outgassed at 623 K for 2 h prior to experiments. Bio-oils samples were obtained by condensation of vapors using a closed flask (60 cm^3) at 253 K using an experimental system shown in Figure 6. Pyrolysis product fractions were quantified gravimetrically and gaseous fraction was calculated by difference.

Taking into consideration the several species identified on bio-oil samples, a solvent fractionation method was used here in order to separate some of wood-extractive derived compounds, considering reported articles [14,41,42]. Bio-oil was diluted with ultrapure water in a ratio of 1:1 (v/v) and centrifuged at 3300 rpm for 5 min in a LW SCIENTIFIC ULTRA 8V apparatus (UCT, Temuco, Chile). Then, water-soluble supernatant was removed and discarded with a syringe. The water-insoluble phase of the bio-oil was added in an SPE cartridge (UCT-CUSIL 500 mg/10 cm^3) in order to carry out a solid extraction stage using a manifold vacuum system. The SPE column was eluted with 10 cm^3 of each solvent in the following sequence: hexane, cyclohexane, diethyl ether, dichloromethane, ethyl acetate, acetone, and acetonitrile. All solvents used here were Lichrosolv grade (Merck, Santiago, Chile).

Eluted samples were analyzed by gas chromatography/mass spectrophotometry (GC/MS) (UCT, Temuco, Chile) in order to identify chemical species, using a Shimadzu QP2010-plus apparatus. GC/MS was configured under the following conditions: an HP5-MS fused silica capillary column with dimensions of 30 m × 0.25 mm × 0.25 mm; an oven temperature starting at 308 K, rising to 453 K at

5 K·min^{-1}, and then up to 573 K at 20 K·min^{-1}; an injector temperature of 523 K with a split ratio of 25:1; helium (Indura 99.999%) as a carrier gas at pressure mode control (10 kPa) and a flow rate of 1 cm^3·min^{-1}; a transfer line and ion source at 523 K; electron energy at 70 eV in SCAN mode (m/z = 35–500 amu). A quadrupole mass detector was operated in an electron impact ionization mode. Data was obtained using GCMSsolution (v2.53) and mass spectra laboratory databases (NIST08 and NIST08s). Computerized matches were manually evaluated by comparing mass spectra and the Kovatz retention index (RI). The Kovatz RI was calculated based on the retention times of the n-alkanes series (C7–C30, Sigma-Aldrich 49451-U, Santiago, Chile) under the same sample chromatographic conditions.

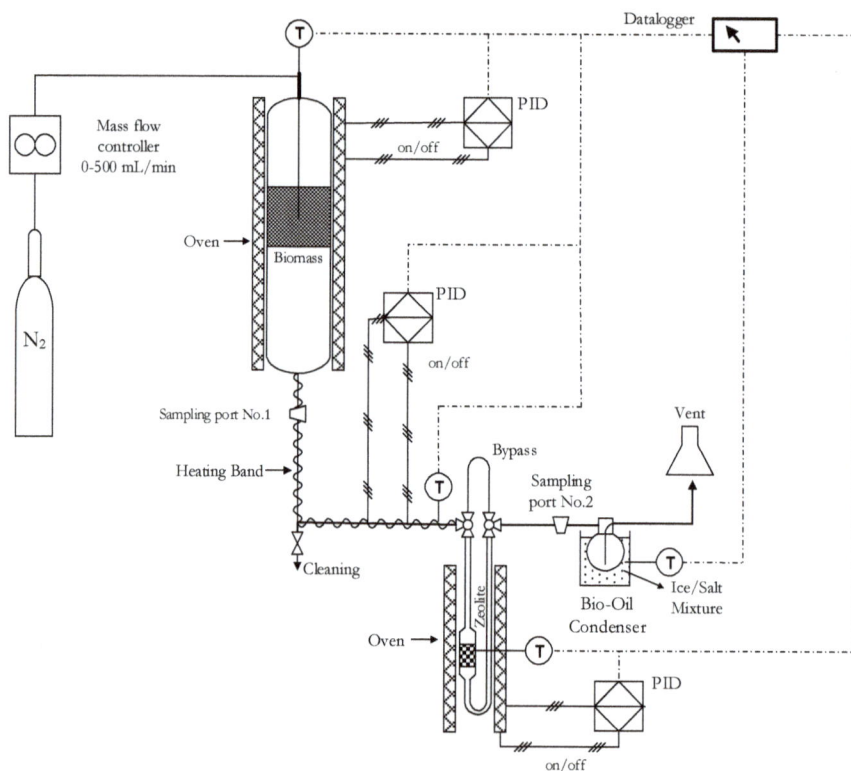

Figure 6. Experimental system.

The viscosity of bio-oil samples was evaluated at different temperatures and storage times. The inclined plane test (IPT) was adopted here, as a quick method to measure kinematic viscosity. A homemade device was developed for this purpose, placing a glass surface in a 45° plane structure. This technique has been previously used for the estimation of viscosity [43]. A calibration curve for the homemade device was obtained using commercial oil samples with known viscosities.

4. Conclusions

To our knowledge, this is the first report of Chilean oak catalytic pyrolysis using Chilean natural zeolite as catalyst. Pyrolysis assays were successfully carried out in the experimental system, designed and made for this study. Brønsted acid sites play a key role on bio-oil upgrade, increasing the bio-oil yield and the composition of hydrocarbons, alcohols, and aldehydes on catalytic bio-oil samples.

The fractionation method developed here becomes crucial to understanding such complex matrices of compounds in oak-derived bio-oils.

Bio-oils are a sustainable source of future energy requirements and value-added chemicals. Thus, Chilean natural zeolites can be considered alternative catalysts that improve bio-oil stability and quality, considering the positive contribution to the viscosity and chemical composition of catalytic bio-oil samples registered in this work.

Acknowledgments: This work was supported by the Chilean National Fund for Scientific and Technological Development (Fondecyt Grant No. 11140781) and the NanoCat$_p$PS research group from University of Bio-Bio, Chile. The authors wish to acknowledge to Scott W. Banks from EBRI, Aston Univ. in Birmingham U.K., for this valuable collaboration on DRIFT experiments.

Author Contributions: S. Alejandro Martín and A. Montecinos conceived and designed the experiments; A. Montecinos performed the experiments; S. Alejandro Martín and C. Cerda-Barrera analyzed the data; S. Alejandro Martín and C. Cerda-Barrera wrote the paper.

Conflicts of Interest: The authors declare no conflict of interest. The founding sponsors had no role in the design of the study; in the collection, analyses, or interpretation of data; in the writing of the manuscript; or in the decision to publish the results.

References

1. Zhao, Y.; Pan, T.; Zuo, Y.; Guo, Q.-X.; Fu, Y. Production of aromatic hydrocarbons through catalytic pyrolysis of 5-hydroxymethylfurfural from biomass. *Bioresour. Technol.* **2013**, *147*, 37–42. [CrossRef] [PubMed]
2. Wyman, C.E.; Dale, B.E.; Elander, R.T.; Holtzapple, M.; Ladisch, M.R.; Lee, Y.Y. Comparative sugar recovery data from laboratory scale application of leading pretreatment technologies to corn stover. *Bioresour. Technol.* **2005**, *96*, 2026–2032. [CrossRef] [PubMed]
3. Demirbas, A.; Arin, G. An overview of biomass pyrolysis. *Energy Source* **2002**, *24*, 471–482. [CrossRef]
4. Zanuttini, M.S.; Lago, C.D.; Querini, C.A.; Peralta, M.A. Deoxygenation of m-cresol on pt/γ-Al$_2$O$_3$ catalysts. *Catal. Today* **2013**, *213*, 9–17. [CrossRef]
5. Messina, L.I.G.; Bonelli, P.R.; Cukierman, A.L. In-situ catalytic pyrolysis of peanut shells using modified natural zeolite. *Fuel Process. Technol.* **2017**, *159*, 160–167. [CrossRef]
6. Mohammed, I.Y.; Kazi, F.K.; Yusup, S.; Alaba, P.A.; Sani, Y.M.; Abakr, Y.A. Catalytic intermediate pyrolysis of napier grass in a fixed bed reactor with ZSM-5, HZSM-5 and zinc-exchanged zeolite-A as the catalyst. *Energies* **2016**, *9*, 246. [CrossRef]
7. Imran, A.; Bramer, E.A.; Seshan, K.; Brem, G. Catalytic flash pyrolysis of biomass using different types of zeolite and online vapor fractionation. *Energies* **2016**, *9*, 187. [CrossRef]
8. Ateş, F.; Putun, A.E.; Putun, E. Fixed bed pyrolysis of euphorbia rigida with different catalysts. *Energy Convers. Manag.* **2005**, *46*, 421–432. [CrossRef]
9. Putun, E.; Uzun, B.B.; Putun, A.E. Fixed-bed catalytic pyrolysis of cotton-seed cake: Effects of pyrolysis temperature, natural zeolite content and sweeping gas flow rate. *Bioresour. Technol.* **2006**, *97*, 701–710. [CrossRef] [PubMed]
10. Aho, A.; Kumar, N.; Eranen, K.; Salmi, T.; Hupa, M.; Murzin, D.Y. Catalytic pyrolysis of woody biomass in a fluidized bed reactor: Influence of the zeolite structure. *Fuel* **2008**, *87*, 2493–2501. [CrossRef]
11. Rajić, N.; Logar, N.Z.; Rečnik, A.; El-Roz, M.; Thibault-Starzyk, F.; Sprenger, P.; Hannevold, L.; Andersen, A.; Stöcker, M. Hardwood lignin pyrolysis in the presence of nano-oxide particles embedded onto natural clinoptilolite. *Microporous Mesoporous Mater.* **2013**, *176*, 162–167. [CrossRef]
12. Corma, A.; Iborra, S.; Velty, A. Chemical routes for the transformation of biomass into chemicals. *Chem. Rev.* **2007**, *107*, 2411–2502. [CrossRef] [PubMed]
13. French, R.; Czernik, S. Catalytic pyrolysis of biomass for biofuels production. *Fuel Process. Technol.* **2010**, *91*, 25–32. [CrossRef]
14. Mohan, D.; Pittman, C.U.; Steele, P.H. Pyrolysis of wood/biomass for bio-oil: A critical review. *Energy Fuels* **2006**, *20*, 848–889. [CrossRef]
15. Alejandro, S.; Valdés, H.; Zaror, C.A. Natural zeolite reactivity towards ozone: The role of acid surface sites. *J. Adv. Oxid. Technol.* **2011**, *14*, 182–189. [CrossRef]

16. Leliveld, B.R.G.; Kerkhoffs, M.J.H.V.; Broersma, F.A.; van Dillen, J.A.J.; Geus, J.W.; Koningsberger, D.C. Acidic properties of synthetic saponites studied by pyridine ir and tpd[ndash]tg of n-propylamine. *J. Chem. Soc. Faraday Trans.* **1998**, *94*, 315–321. [CrossRef]

17. Konan, K.L.; Peyratout, C.; Smith, A.; Bonnet, J.P.; Magnoux, P.; Ayrault, P. Surface modifications of illite in concentrated lime solutions investigated by pyridine adsorption. *J. Colloid Interface Sci.* **2012**, *382*, 17–21. [CrossRef] [PubMed]

18. Simon-Masseron, A.; Marques, J.P.; Lopes, J.M.; Ribeiro, F.R.A.; Gener, I.; Guisnet, M. Influence of the si/al ratio and crystal size on the acidity and activity of hbea zeolites. *Appl. Catal. A Gen.* **2007**, *316*, 75–82. [CrossRef]

19. Lee, H.W.; Kim, Y.-M.; Lee, B.; Kim, S.; Jae, J.; Jung, S.-C.; Kim, T.-W.; Park, Y.-K. Catalytic copyrolysis of torrefied cork oak and high density polyethylene over a mesoporous hy catalyst. *Catal. Today* **2017**, in press. [CrossRef]

20. Chen, D.; Zhou, J.; Zhang, Q.; Zhu, X. Evaluation methods and research progresses in bio-oil storage stability. *Renew. Sustain. Energy Rev.* **2014**, *40*, 69–79. [CrossRef]

21. Ba, T.; Chaala, A.; Garcia-Perez, M.; Roy, C. Colloidal properties of bio-oils obtained by vacuum pyrolysis of softwood bark. Storage stability. *Energy Fuels* **2004**, *18*, 188–201. [CrossRef]

22. Holladay, J.E.; White, J.F.; Bozell, J.J.; Johnson, D. *Top Value Added Chemicals from Biomass—Volume II, Results of Screening for Potential Candidates from Biorefinery Lignin*; Pacific Northwest National Lab. (PNNL): Richland, WA, USA; National Renewable Energy Laboratory (NREL): Golden, CO, USA, 2007.

23. Imam, T.; Capareda, S. Characterization of bio-oil, syn-gas and bio-char from switchgrass pyrolysis at various temperatures. *J. Anal. Appl. Pyrolysis* **2012**, *93*, 170–177. [CrossRef]

24. Cardoso, C.A.L.; Machado, M.E.; Caramão, E.B. Characterization of bio-oils obtained from pyrolysis of bocaiuva residues. *Renew. Energy* **2016**, *91*, 21–31. [CrossRef]

25. Lazzari, E.; Schena, T.; Primaz, C.T.; da Silva Maciel, G.P.; Machado, M.E.; Cardoso, C.A.L.; Jacques, R.A.; Caramão, E.B. Production and chromatographic characterization of bio-oil from the pyrolysis of mango seed waste. *Ind. Crops Prod.* **2016**, *83*, 529–536. [CrossRef]

26. Zhang, Y.; Chen, P.; Lou, H. In situ catalytic conversion of biomass fast pyrolysis vapors on HZSM-5. *J. Energy Chem.* **2016**, *25*, 427–433. [CrossRef]

27. Mihalcik, D.J.; Mullen, C.A.; Boateng, A.A. Screening acidic zeolites for catalytic fast pyrolysis of biomass and its components. *J. Anal. Appl. Pyrolysis* **2011**, *92*, 224–232. [CrossRef]

28. Zanuttini, M.S.; Peralta, M.A.; Querini, C.A. Deoxygenation of m-cresol: Deactivation and regeneration of pt/γ-Al$_2$O$_3$ catalysts. *Ind. Eng. Chem. Res.* **2015**, *54*, 4929–4939. [CrossRef]

29. Aho, A.; Kumar, N.; Eranen, K.; Salmi, T.; Hupa, M.; Murzin, D.Y. Catalytic pyrolysis of biomass in a fluidized bed reactor: Influence of the acidity of h-beta zeolite. *Process Saf. Environ. Prot.* **2007**, *85*, 473–480. [CrossRef]

30. Zhang, Q.; Chang, J.; Wang, T.; Xu, Y. Review of biomass pyrolysis oil properties and upgrading research. *Energy Convers. Manag.* **2007**, *48*, 87–92. [CrossRef]

31. Sipilä, K.; Kuoppala, E.; Fagernäs, L.; Oasmaa, A. Characterization of biomass-based flash pyrolysis oils. *Biomass Bioenergy* **1998**, *14*, 103–113. [CrossRef]

32. Diebold, J.P. *A Review of the Chemical and Physical Mechanisms of the Storage Stability of Fast Pyrolysis Bio-Oils*; National Renewable Energy Lab.: Golden, CO, USA, 1999.

33. Boucher, M.; Chaala, A.; Roy, C. Bio-oils obtained by vacuum pyrolysis of softwood bark as a liquid fuel for gas turbines. Part I: Properties of bio-oil and its blends with methanol and a pyrolytic aqueous phase. *Biomass Bioenergy* **2000**, *19*, 337–350. [CrossRef]

34. Yu, F.; Deng, S.; Chen, P.; Liu, Y.; Wan, Y.; Olson, A.; Kittelson, D.; Ruan, R. Physical and chemical properties of bio-oils from microwave pyrolysis of corn stover. *Appl. Biochem. Biotechnol.* **2007**, *137*, 957–970. [PubMed]

35. Adam, J.; Blazsó, M.; Mészáros, E.; Stocker, M.; Nilsen, M.H.; Bouzga, A.; Hustad, J.E.; Gronli, M.; Oye, G. Pyrolysis of biomass in the presence of al-mcm-41 type catalysts. *Fuel* **2005**, *84*, 1494–1502. [CrossRef]

36. Wang, S.; Gu, Y.; Liu, Q.; Yao, Y.; Guo, Z.; Luo, Z.; Cen, K. Separation of bio-oil by molecular distillation. *Fuel Process. Technol.* **2009**, *90*, 738–745. [CrossRef]

37. Erol, M.; Haykiri-Acma, H.; Küçükbayrak, S. Calorific value estimation of biomass from their proximate analyses data. *Renew. Energy* **2010**, *35*, 170–173. [CrossRef]

38. Parikh, J.; Channiwala, S.A.; Ghosal, G.K. A correlation for calculating hhv from proximate analysis of solid fuels. *Fuel* **2005**, *84*, 487–494. [CrossRef]
39. Valdés, H.; Farfán, V.J.; Manoli, J.A.; Zaror, C.A. Catalytic ozone aqueous decomposition promoted by natural zeolite and volcanic sand. *J. Hazard. Mater.* **2009**, *165*, 915–922. [CrossRef] [PubMed]
40. Valdés, H.; Alejandro, S.; Zaror, C.A. Natural zeolite reactivity towards ozone: The role of compensating cations. *J. Hazard. Mater.* **2012**, *227–228*, 34–40. [CrossRef] [PubMed]
41. Garcia-Perez, M.; Chaala, A.; Pakdel, H.; Kretschmer, D.; Roy, C. Characterization of bio-oils in chemical families. *Biomass Bioenergy* **2007**, *31*, 222–242. [CrossRef]
42. Kanaujia, P.K.; Sharma, Y.K.; Garg, M.O.; Tripathi, D.; Singh, R. Review of analytical strategies in the production and upgrading of bio-oils derived from lignocellulosic biomass. *J. Anal. Appl. Pyrolysis* **2014**, *105*, 55–74. [CrossRef]
43. Dey, A.; Riaz, S. Viscosity measurement of mould fluxes using inclined plane test and development of mathematical model. *Ironmak. Steelmak.* **2012**, *39*, 391–397. [CrossRef]

catalysts

MDPI

Review

Catalytic Hydroisomerization Upgrading of Vegetable Oil-Based Insulating Oil

Dieu-Phuong Phan and Eun Yeol Lee *

Department of Chemical Engineering, Kyung Hee University, Gyeonggi-do 17104, Korea;
phanphuong07h5@gmail.com
* Correspondence: eunylee@khu.ac.kr; Tel.: +82-31-201-3839

Received: 9 February 2018; Accepted: 28 March 2018; Published: 28 March 2018

Abstract: Due to its high biodegradability, high dielectric strength, and good thermal stability, vegetable oil is under consideration as an alternative transformer fluid for power system equipment, replacing traditional petroleum-based insulating oils. Its main drawbacks are its poor low-temperature properties arising from the crystallization of its long-chain normal paraffins, and its lower oxidative stability arising from its higher concentration of unsaturated fatty acids. Hydroisomerization/isomerization over bifunctional catalysts is considered to be an efficient pathway to upgrade vegetable oil-based insulating oil; this converts saturated/unsaturated long-chain fatty acids to branched isomers. The efficiency of this process depends crucially on the behavior of the catalyst system. This paper extensively reviews recent results on the influence that the metal phase and acidity, the effects of pore channels, and the balance between metal and acid sites have upon the activity and selectivity of catalytic hydroisomerization.

Keywords: insulating oils; hydroisomerization; bio-insulating oil; zeolite

1. Introduction

1.1. Importance of Biologically Sourced Insulating Oil

Insulating oils are liquid dielectrics that are extensively used in transformers, one of the most important pieces of equipment in the power system. The important functions of transformer oil are to ensure electrical insulation and to act as a cooling fluid, removing heat from the windings and the core [1]; thus, the oil determines the life span of the equipment. Petroleum-based oils, also called mineral oils, have been used as transformer oils since the 1880s due to their good compatibility with cellulose paper insulation, suitable physicochemical and heat dissipation properties, and low cost [2]. However, these oils have many negative aspects including their non-renewability, low flash point, and low fire point, as well as the various environmental, political, and socioeconomic issues associated with petroleum products. Polychlorinated aromatic hydrocarbon fluids (such as Polychlorinated Biphenyls—PCBs), silicones, and synthetic ester fluids have been used as alternative fluids thanks to their outstanding properties such as higher fire points compared to mineral oil and good oxidative and thermal stabilities. However, their availability and comparatively higher cost have confined their application to special transformers [3,4]. Furthermore, these oils are poorly biodegradable, flammable, and have negative environmental impacts, since equipment failure and leakage can contaminate soil and water.

Vegetable oils, which are natural esters, are considered to be promising candidates for insulating oils owing to their high biodegradability, good thermal properties, and wide availability [5,6]. The use of vegetable oil-based insulating oil could resolve environmental problems thanks to their high biodegradability and lack of toxicity in nature. Moreover, vegetable oils are less expensive than synthetic esters, and in the long run may be even cheaper than mineral oil. In addition, owing to their hydrophilicity, vegetable-based insulating oil could increase the life of transformers compared to conventional mineral oils (Table 1).

Table 1. Properties of typical transformer dielectric fluids (modified from Reference [7]).

	Mineral Oils	Silicone Oils	Synthetic Esters	Vegetable Oils
Dielectric breakdown, kV—ASTM D1816	45–85	35–60	43–70	49–97
Kinematic viscosity at 0 °C, centistokes—cSt	<76	81–92	26–240	77–500
at 40 °C	3–16	35–40	14–29	16–50
at 100 °C	2–3	15–17	4–6	4–15
Pour point, °C—ASTM D97	−30 to −60	−50 to −60	−40 to −60	−10 to −33
Flash point, °C—ASTM D92	110–175	300–310	250–310	310–343
Fire point, °C—ASTM D92	110–185	340–350	300–310	350–360
Density at 20 °C, kg·dm^3—ASTM D1298	0.83–0.89	0.96–1.10	0.90–1.00	0.87–0.92
Biodegradability over 21 days—CEC-L-33	<30%	0%	80%	97%–99%
Specific heat, J·g^{-1}·k^{-1}	1.6–2.0	1.5	1.8–2.3	1.5–2.1
Water solubility, ppm, at 20 °C	45	200	2700	
at 100 °C	650	1100	7200	
Expansion coefficient, 10^{-4} k^{-1}	7–9	10	6.5–10	5.5–5.9

The main fatty esters present in biologically sourced insulating oil (hereinafter 'vegetable oil-based insulating oil') are palmitic acid (C16:0), stearic acid (18:0), oleic acid (18:1), linoleic acid (18:2), and linolenic acid (18:3). This applies to oils from various biomass feedstocks, as listed in Table 2, including soybean, sunflower, rapeseed (i.e., canola), palm, and peanut oils. Various other fatty acids are present in minor amounts in virtually all oils and fats derived from bio-oil feedstocks.

Table 2. Fatty acid compositions of various vegetable oils (modified from Reference [8]).

	Fatty Acid		Microalgae (Sp)	Palm	Olive	Peanut	Rape-Seed	Soybean	Sunflower	Grape	Almond	Corn	Coconut
Saturated	Capric	C10:0	–	–	–	–	–	–	–	–	–	–	6.0
	Lauric	C12:0	–	0.1	–	0.1	–	–	–	–	–	–	47.0
	Myristic	C14:0	0.3	0.7	–	0.1	–	–	–	0.1	–	–	18.0
	Palmitic	C16:0	40.2	36.7	11.6	8.0	4.9	11.3	6.2	4.6	10.4	6.5	9.0
	Stearic	C18:0	1.2	6.6	3.1	1.8	1.6	3.6	3.7	3.4	2.9	1.4	3.0
	Arachidic	C20:0	0.1	0.4	0.03	0.9	–	0.3	0.3	0.3	0.3	0.1	–
	Behenic	C22:0	–	0.1	0.1	3	–	–	0.7	0.7	0.1	–	–
Mono-unsaturated	Palmitoleic	C16:1	9.2	–	–	–	–	–	–	–	–	–	–
	Oleic	C18:1	5.4	46.1	75.0	53.3	33.0	24.9	25.2	62.8	77.1	65.6	6.0
	Gadoleic	C20:1	–	0.2	–	2.4	9.3	0.3	0.2	–	–	0.1	–
	Erucic	C22:1	–	–	–	–	23.0	0.3	0.1	–	–	0.1	–
Poly-unsaturated	Linoleic	C18:2	17.9	8.6	7.8	28.4	20.4	53.0	63.1	27.5	7.6	25.2	2.0
	Linolenic	C18:3	18.3	0.3	0.6	0.3	7.9	6.1	0.2	0.1	0.8	0.1	–

Sp: *Spirulina maxima*.

1.2. Application Drawbacks of Vegetable Oil-Based Insulating Oils

Although vegetable oil-based insulating oils have various advantages as discussed above, they also have the drawbacks of high pour points, higher viscosity, and poorer aging compared to conventional mineral oils [9–11]. For instance, the pour points of common vegetable esters are in the range from −10 to −33 °C due to their higher contents of long-chain normal paraffins compared to mineral oils, which can be a critical problem for low-temperature applications [12]. However, the structural properties of acid chain length, unsaturation, and branching can be altered to tailor the pour points of vegetable oil-based insulating oils.

Secondly, under normal transformer operating temperatures, vegetable oil-based insulating oils show higher viscosities than mineral oils, meaning that they will yield less efficient convection heat transfer and thus higher operation temperatures. Moreover, viscosity strongly influences the impregnation of cellulose-based solid insulation. It has been observed that with natural esters it takes at least two times as long to complete the impregnation of laminated pressboard. This process can be hastened by carrying it out at a slightly higher temperature [13].

Liquid insulators used in transformers must show higher stabilities toward oxidation. Oxidation is the most influential factor in oil aging, and is especially important for oils used in free-breathing transformers [11]. Actually, the oxidative stabilities of natural esters depend on their fatty acid distributions, the refining processes used, and the presence of natural antioxidants. It is known that increasing the unsaturated fatty acid content decreases the oxidative stability of natural esters. For example, the oxidation stability of oleic acid (having one double bond) is 10 times that of linoleic acid (having two double bonds), which is in turn twice that of linolenic acid (having three double bonds).

Solving these problems simultaneously has been challenging because solving one problem often worsens another. The properties of vegetable oil-based insulating oils depend mainly on the structures of their constituent fatty acids. Hence, the properties of these esters can be tailored by choosing a suitable ratio of saturated to unsaturated triglycerides during their formulation. Branched-chain saturated fatty acids are ideal components because they have better low-temperature properties than linear saturated fatty acid and show greater oxidative stability than linear unsaturated fatty acids [14].

The aim of this review is to summarize recent processes to produce high-quality insulating oils from renewable sources. We focus on catalytic processes to transform the normal long-chain hydrocarbons of fatty acids to branched isomers that are desirable components for vegetable oil-based insulating oils and bio-fuel oils in general.

2. Catalytic Hydroisomerization/Isomerization of Long-Chain Saturated/Unsaturated Fatty Acids

2.1. Hydroisomerization/Isomerization for Upgrading Vegetable Oil-Based Insulating Oils

In the pathway of vegetable oil-based insulating oil upgrading, catalytic hydroisomerization is a more sustainable process for reducing the concentration of linear paraffins by transforming them to their branched isomers. According to Abhari et al. [15], the pour point of vegetable oil was decreased to −10 °C by applying the hydroisomerization over Pt/Pd on an amorphous silica/alumina catalyst. The hydroisomerized product had the *iso*-paraffin to *n*-paraffin ratio in the range of 1:1 to 10:1, of which about 80% were mono-methyl branched paraffins. Reaume et al. [16,17] showed that the synergetic use of hydroisomerization and isomerization can decrease the cloud points (defined as the temperature at which the oil becomes cloudy) of oleic methyl ester, palmitic methyl ester, and various natural esters by 7.5 to 12.9 °C. Isomerization is used to convert unsaturated fatty acids already having C=C bonds, whereas hydroisomerization is used to introduce these π bonds into saturated chains (creating a carbon-carbon double bond site), which requires a dehydrogenation/hydrogenation step. Firstly, the dehydrogenation of normal paraffins takes place over a metal phase. The generated olefins diffuse and protonate on the Brønsted acid sites to form alkylcarbenium ions, which then undergo skeletal isomerization and then deprotonation by means of beta scission. After desorbing from acid sites and diffusing to metal sites, these isomerized olefins are then hydrogenated to form the corresponding paraffins [18]. The desired and undesired products of these reactions are shown in Figure 1. These two different reactions are thus indispensable to improve the low-temperature properties of insulating and lubricant oils.

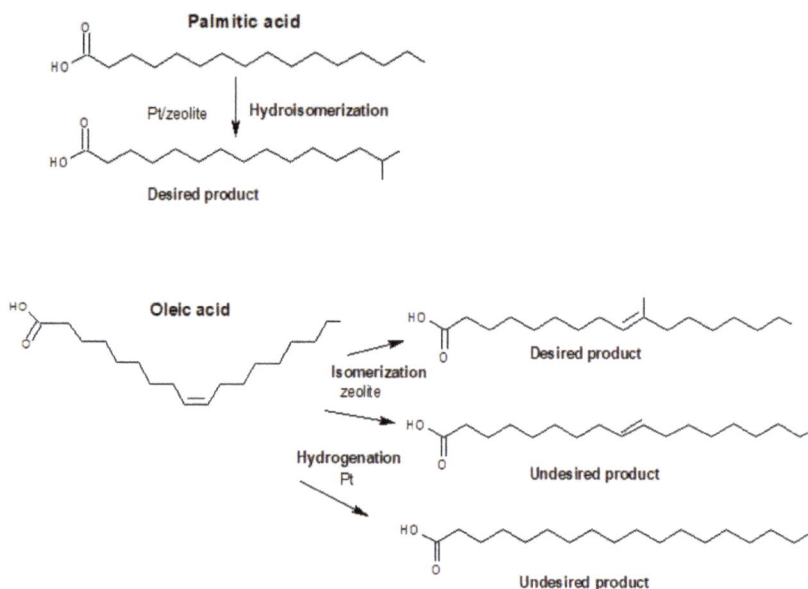

Figure 1. Desired and undesired products from the isomerization/hydroisomerization of oleic acid and the hydroisomerization of palmitic acid (modified from Reference [16]).

Because they allow a combination of dehydrogenation/hydrogenation and acid catalysis that is highly desirable for bio-oil upgrading, bifunctional metal/acid catalysts have attracted the attention of many researchers worldwide, who have explored their special catalytic performance and worked to elucidate their mechanisms [16,17,19–26]. Various catalysts have been reported for *n*-paraffin transformation, with a wide range of performance depending upon reaction temperature, metal site types, acidity, support structure effects, and the balance between metal and acid sites. Generally, an effective bifunctional catalyst for hydroisomerization should diminish side reactions such as hydrocracking and oligomerization. The decomposition of carbocations during hydrocracking occurs by means of beta scission of C–C bonds, the selectivity of which depends on the natural and structural effects of acid sites in the catalyst support. According to Coonradt and Garwood [27], these differences in selectivity arise from catalysts' differences in acidity and relative strengths of dehydrogenation/hydrogenation activity. The balance between metal and Brønsted acid sites plays a vital role in determining the selectivity of the catalyst; high metal site density limits the formation of multiply branched isomers by promoting hydrogenation of the primary isomerization products, an issue that has received relatively little attention, with only a few reports so far [28]. Metal dispersion, metal loading, acidity and density of acid sites, and porous channel system effect are also key properties for determining the catalyst selection and experimental design [29–32].

2.2. Catalyst System for Hydroisomerization/Isomerization

2.2.1. Metallic Function

The metal phase type is important to the catalytic performance. Both noble metals (Pt, Pd, etc.) and non-noble metal catalysts (Ni, Co, Cu, etc.) have been found to be efficient metallic catalysts for normal long-chain paraffin hydroisomerization [22]. Compared to non-noble metal types, noble metal catalysts show considerably higher activity and selectivity for hydroisomerization due to their easy activation of hydrogen, which then spills over onto the surface of support [33]. Zhang et al. [21] evaluated the role of the metal phase upon the catalytic performance of tungstenated zirconia for *n*-hexadecane transformation; among the different metals they evaluated (Pd, Pt, and Ni), with varying metal loadings in one case, platinum was found to be the best for the metallic phase. Metal content is also a key factor for tailoring the catalytic performance of supported metal catalysts. Song et al. [30] investigated the effect of Pd loading upon the catalytic performance of *n*-decane hydroisomerization over a Pd-based catalyst supported on SAPO-11 zeolite in a fixed-bed continuous microreactor. The metal dispersion on SAPO-11 decreased with increasing Pd loading, owing to the increasing agglomeration of metal particles on the support surface. Additionally, increasing the metal particle size can partially block some pore mouths of SAPO-11, impairing its function as a molecular sieve and thereby dramatically increasing the ratio between the density of accessible Pd metal sites and the number of available Brønsted acid sites. For *n*-decane hydroisomerization, the activity of xPd/SAPO-11 catalyst decreases in order of decreasing metal dispersion according to the following trend: 0.1Pd/SAPO-11 > 0.3Pd/SAPO-11 > 0.05Pd/SAPO-11 > 0.5Pd/SAPO-11. The lower activity of the catalyst, having the lowest Pd content, is due to its deficiency of metal sites; in this case of low Pd loading, the rate of the transformation is limited by the rate of hydrogenation/dehydrogenation over the Pd sites.

Frank Bauer et al. [34] focused on the hydroisomerization of *n*-hexadecane over nanosized bimetallic Pt-Pd/H-Beta catalysts, finding that the Pt/H-beta catalyst is significantly more selective for hydroisomerization than its Pd-containing counterparts, owing to the higher hydrogenating activity of Pt relative to Pd. On the other hand, the metal dispersion and bimetallic interaction play important roles in the hydroisomerization activity. In fact, Bauer et al. found that the branched isomer selectivity was superior to 90% at the conversion of 64.8% over the Pt/H-beta catalyst. With gradual increasing in the loading of Pd as a promoter, the hydroisomerization activity remarkably increased due to increasing noble metal dispersion, specifically as a result of increasing the concentration of the easily reducible palladium oxides present after calcination, which promote hydroisomerization [34,35] (Table 3).

Table 3. Metal dispersion, *n*-hexadecane conversion, and isomerized selectivity/cracked selectivities of Pd-Pt/H-Beta catalysts with various Pt and Pd loadings. Reaction conditions: T = 200 °C, P = 1 bar, liquid hourly space velocity (LHSV) = 3 h^{-1} (modified from Reference [34]).

H-Beta Based Catalyst with Pt-Pd Loading (wt %/wt %)	Pd-Pt (0.6/0.0)	Pd-Pt (0.6/0.2)	Pd-Pt (0.4/0.4)	Pd-Pt (0.2/0.6)	Pd-Pt (0.0/0.8)
Metal dispersion (%)	53	92	91	77	54
Conversion (%)	54.9	77.0	77.6	74.2	64.8
Isomerized selectivity (%)	80.4	91.2	92.9	92.3	93.4
Cracked selectivity (%)	19.6	8.8	7.1	7.7	6.2

A. de Lucas et al. [22] also focused on the influence of Pt-Pd metal loading and how the combination of these metals supported on catalysts based on agglomerated beta zeolite performed in the hydroisomerization of *n*-octane. For equal loadings, the metallic dispersion was higher in the case of platinum-based catalysts compared to palladium-based catalysts because their greater polarizability—due to the d electron configuration and larger atomic size—make Pt atoms less mobile than Pd atoms [36]. However, in the case of bimetallic Pt-Pd supported on agglomerated beta zeolite, the metal dispersion decreases with increasing palladium content owing to the interaction between palladium and platinum in bimetallic particles, as K. Lee et al. [37] and Pawelec et al. [38] demonstrated.

2.2.2. Acid Zeolites

In recent decades, many studies have been performed on the hydroisomerization of long-chain paraffins over zeolite catalysts thanks to their suitable acidity or hierarchical structures that combine microporous and mesoporous channel systems. In fact, due to their peculiar pore channel, zeolite-based catalysts are used to improve the degree of branching of product and decrease the tendency toward cracking reaction through the transition state selectivity (TSS) and product shape selectivity (PSS) effects. On the other hand, the acidity of the zeolite greatly influences the hydroisomerized yield. The acid strength distribution and acid site density both influence this yield, and the relation between metal sites and protonating sites is critical in determining the performance of the bifunctional catalysts. Many authors have concluded that weak or medium-strength Brønsted acid sites are responsible for promoting isomerized selectivity, whereas strong Brønsted acid and Lewis acid sites tend to promote cracking [19,20,25,29,39,40]. Hence, modifying the Brønsted acidity of the zeolite is essential to improve its hydroisomerization selectivity.

Acid Sites

The acidity of the zeolite is a crucial factor in determining the overall catalytic performance in the upgrading of vegetable oils to produce low-pour-point insulating and lubricating oils. The degree of acidity and the acid site types (Brønsted or Lewis) determine the degree to which the cracking or isomerization reaction is favored. Mériaudeau et al. [40] studied the performance of 1%Pt/SAPO-41, 1%Pt/SAPO-31, and 1%Pt/SAPO-11 catalysts in *n*-octane hydroisomerization through a fixed-bed continuous flow reactor at 200–400 °C under atmospheric pressure. The catalytic activity decreased in increasing order of the amount and strength of acid sites: SAPO-41 > SAPO-11 > SAPO-31. However, the reaction rate was most strongly restricted by diffusion in the microporous channels of the SAPO-type zeolite. The low activity of the catalyst based on SAPO-31 zeolite resulted from the combined effect of its diffusion limitations and low acidity. Verma et al. [39] synthesized hierarchical mesoporous crystalline HZSM-5 zeolite catalysts in high-surface-area and low-surface-area types for the hydroconversion of algae and Jatropha seed oil to jet fuel. According to ^{27}Al MAS NMR analysis, the concentration of extra-framework Al present in the high-surface-area zeolite sample was higher than that in its low-surface-area counterpart, representing an increased number of Lewis acid sites in the former [41]. The former catalyst type also had more strong acid sites, favoring cracking and thus decreasing its isomerization selectivity. SAPO-11 and ZSM-22 have been used widely in paraffin long-chain hydroisomerization catalysts because of their moderate acidity and straight channels of 10-membered rings. The drawback of this zeolite type is its lower acid site density at the pore mouths where the main isomerization occurs, according to the pore mouth and key-lock mechanism [42,43].

The acidity of zeolite can be also controlled by modifying its Si/Al ratio by means of ion exchange or post-synthesis dealumination or desilication treatments, or by creating a siliceous border over the pore mouth of the channel system. The advantages of these tailoring methods are that they change both the total number of acidic sites and the density of electrons on the linking hydroxyl group, thereby changing the Brønsted acidity. Parma et al. [28] studied the effect of the zeolite Si/Al ratio in the Pt/ZSM-22 catalyst system upon the branched isomer selectivity for *n*-hexadecane hydroisomerization (Table 4). Decreases in the total number of acid sites and in the number of Brønsted acid sites correspond to increases in the Si/Al framework ratio. At the constant conversion level of 90%, a catalyst having a lower Si/Al ratio showed excellent selectivity and maximum isomer yield at lower reaction temperatures, 300–320 °C, compared to the reaction temperature range of 330–350 °C, providing maximum isomer yield for a catalyst with a higher Si/Al ratio. This result can be attributed to the mild Brønsted acidic strength of ZSM-22 zeolite, which favors isomerization over cracking at lower temperatures. Furthermore, relative to CAT-1 (having the Si/Al ratio of 30), CAT-2 (having the

Si/Al ratio of 45) has a lower reaction temperature that maximizes isomer yield, indicating that it has optimal acid function and metal site balance over the mouths of its zeolite pores.

Table 4. Acidity properties of H-ZSM-22 catalysts with various Si/Al ratios and their reaction temperatures for obtaining 90% conversion (modified from Reference [28]).

Catalyst	Bulk Si/Al Ratio	Total Acid Sites (μmol NH₃/g)	Brønsted Acid Sites (μmol NH₃/g)	Lewis Acid Sites (μmol NH₃/g)	Temperature Giving 90% Conversion (°C)
CAT-1	30	217.5	177.1	40.4	320
CAT-2	45	146.6	127.5	19.1	305
CAT-3	60	115.0	104.3	10.7	330
CAT-4	90	78.8	71.3	7.5	350

However, many authors have also stated that the number of acid centers of this zeolite type and their strength mainly determine their activity as hydroisomerization catalysts while having virtually no effect upon their selectivity, which instead depends more upon the structure and behavior of their pore systems.

Pore Structure Effect

Molecular sieve-based catalysts are used widely for hydroisomerization because of their narrow pores and the restricted access to and escape from their inner surfaces; the resulting behavior is termed pore mouth and key-lock catalysis [44–46]. The effects of the pore channels upon the metal/zeolite catalyst performance and the product distribution have been studied in depth in various efforts. The ZSM-22 zeolite, having TON (Theta-One)-framework topology, and SAPO-11, having the AEL (Aluminophosphates with sequence number ELeven)-framework structure consisting of one-dimensional channels each comprised of 10-membered rings, have been studied extensively for hydroisomerization [19,25,29,30,32,45,47–50]. Martens et al. [43,45,46,51] demonstrated that the specific confined space channels of these two zeolite structures allow their peculiar methyl branching selectivity and hinder decomposition reactions inside the pores. These authors studied the formations of monomethyl and dimethyl branching groups of long-chain paraffins from decane to tetracosane during hydroisomerization over the Pt/ZSM-22-bifunctional catalyst, based on the pore mouth and key-lock mechanism [46]. The maximal isomers yields were 77%–90%, with the increase of mono- and multibranched isomer yields depending on the length of the paraffin chain. In detail, the mono- and multibranched isomer yields are 55% and 22% for *n*-C10 transformation, respectively, compared to over 80% and 70% in the case of higher carbon number paraffin chains as the initial material (*n*-C20, *n*-C22, and *n*-C24). Nghiem et al. [52] investigated Pt-based catalysts supported on ZSM and SAPO zeolite types with one-dimensional tubular and non-intersection medium pore structure for *n*-octane hydroisomerization. The catalyst performance and selectivity based on these zeolite types are shown in Table 5.

At 15% conversion, each of the listed catalysts has an isomerized selectivity of over 98%, excepting those with large pore structures, ZSM-12 and SAPO-5, with 89% and 79% selectivities. Moreover, the maximum isomerized yields versus *n*-octane conversion are only 50% for ZSM-12 and less than 30% for SAPO-5, whereas the other molecular sieves in this studied series gave 77–81 wt % isomerized yields.

Chi Kebin et al. [53] investigated the performance of platinum supported on ZSM-22/ZSM-23 zeolite mixtures as hydroisomerization catalysts. The ZSM-22/ZSM-23 zeolites were prepared to have a uniform needle-shaped particle size from spindle-shaped ZSM-22 and nest-shaped ZSM-23. These catalysts displayed unique molecular shape selectivity with a pore cross-section of 0.45 × 0.55 nm and had similar physical properties. At low conversions, the principal hydroisomerized products were monobranched isomers for all catalysts. This primary product transformed to multibranched isomers as secondary products when the degree of conversion was above 80%, due to competitive adsorption on the catalyst surface favoring *n*-alkanes over monobranched isomers. In order of decreasing acidity, namely ZSM-23 > ZSM-22 > ZSM-22/ZSM-23, the hydrocracked products were decreasingly predominant at high conversions (>85%), with >35%, 27%, and 19% yields over the respective supported platinum catalysts. However, as is common to all 10-membered ring zeolites, a drawback of these catalysts is that they hinder the hydrocracking reaction, thereby also inhibiting the generation of multibranched isomers due to increasing product diffusion limitations with increasing paraffin length. Actually, the hydroisomerization reaction would take place at the pore mouth while the carbon bears the positive charge of the stable carbenium ion localized inside the pore. This means that only the external surface of a ZSM-22 or SAPO-11 crystal contributes to the catalytic activity, whereas the remaining part of the crystal is catalytically inactive. Moreover, the diffusion limitations and confined access may also result in micropore blockage by large molecules or catalyst deactivation by means of coke deposition. Therefore, tailoring the pore architecture of zeolite-based catalysts to solve their diffusion limitations is highly desirable.

Table 5. Isomer selectivities of SAPO and ZSM zeolite types for *n*-octane hydroisomerization. Reaction conditions: T_R = 250 °C, H_2/HC = 60, conversion α = 15%, thermodynamic value 2-MeC_7/3-MeC_7 = 0.90 (modified from Reference [52]).

Type of Zeolite	Dimensionality	Description	Isomer Selectivity (wt %) at Conversion α = 15%	Maximum Isomer Yields (wt %)
SAPO-5	1-D	12-membered ring (12-R) channels each with pore opening 0.7 × 0.7 nm	79	<30
SAPO-31	1-D	12-membered ring (12-R) channels each of 0.54 × 0.54 nm diameter	99	78
SAPO-11	1-D	10-R channels each with an elliptical pore opening 0.39 × 0.64 nm	98	78
SAPO-41	1-D	Elliptical 10-R channels each 0.43 × 0.70 in diameter	99.5	81
ZSM-12	1-D	12-R channels each with pore diameter 0.55 × 0.62 nm	89	50
ZSM-48	1-D	10-R channels each with pore diameter 0.53 × 0.56 nm	99	77
ZSM-22	1-D	10-R channels each with pore diameter 0.45 × 0.55 nm	99.4	81

Vandegehuchte et al. [54] investigated the effect of mixing a ZSM-22 zeolite (Si/Al = 45) with a non-shape-selective Y zeolite (Si/Al = 2.6) in a Pt-based catalyst for *n*-C_{10} to C_{13} chain hydroisomerization via single-event microkinetic model simulations. The aim of this research was to optimize the synergy between primary monobranching on the more active ZSM-22 and secondary dibranching on Y. The best performance was obtained by using a zeolite mixture containing 75% ZSM-22. The maximal isomerized yields in this case reached 80% for *n*-C_{10} as the initial feed and 63% for a commercial mixture of *n*-C_{10} to C_{13} paraffin at the conversion level of 90%.

A series of Pt/ZSM-22 catalysts with various siliceous degrees were synthesized for *n*-dodecane isomerization by Niu et al. [29]. Their work showed that the highly siliceous ZSM-22-based catalyst performed better than silica-alumina ZSM-22 analogues in terms of isomerized selectivity and hindering of the cracking reaction. The effect of Brønsted acidity upon isomerization was investigated by introducing acidic silanol groups. It was demonstrated that the acidic silanol group in pure silica zeolite is consider as a second kind of acid site, thereby improving the activity and isomer selectivity of the ZSM-22-based hydroisomerization catalyst when the amount of typical Brønsted acid sites is inadequate. In fact, the catalysis selectivity slightly increased with the Si/Al ratio; also, the selectivity of a catalyst based on siliceous ZSM-22 was higher than that based on a silica-alumina mixture by 5%–7% over conversion, reaching the highest isomerized selectivity of 95.7% and maximum yield of 80.2%.

In recent years, zeolites exhibiting hierarchical porosity, with their inherent microporous system and additional mesoporosity, have been investigated as alternative solutions for the abovementioned problems [32,55,56]. Mesoporosity can be induced by introducing intercrystalline mesopores into the nanoscale zeolite crystals or by creating a system of intracrystalline mesopores in the microporous channels (Figure 2). The advantage of such hierarchical systems is that they can integrate the shape selectivity of the intracrystalline micropores and the efficacious mass transfer of the mesoporous system because of their increased diffusivity and decreased diffusion path length [47]. Martens et al. [45] studied the impact of hierarchical ZSM-22-based catalyst upon the hydroisomerization reaction pathways of the *n*-decane, *n*-nonadecane, and pristane model molecules. Both conventional and hierarchized zeolites had the same platinum content and similar dispersion behavior. The efficiency of hierarchization is demonstrated by the increase of maximum isomer yields of *n*-decane hydroisomerization, reaching 82% versus the 67% of the conventional Pt/ZSM-22 catalyst. Similarly, in the case of *n*-nonadecane, the conventional ZSM-22-based catalyst showed a relatively high maximum yield of 88%, which was further improved to 92% by hierarchization. Particularly, the formation of multibranched isomers was also enhanced. As mentioned above, the selectivity for multibranched isomers over this zeolite type was hindered; this selectivity was improved considerably by hierarchization. The maximum multibranch yields of 35% and 70% were observed at extremely high conversion levels in the cases of *n*-decane and *n*-nonadecane, respectively.

Furthermore, the greatest benefit of this hierarchical zeolite is its tendency to produce branched isomers having methyl groups near the center of the carbon chain; these are the most promising components to impart the desired low-temperature properties to vegetable oil-based insulating oils (Table 6).

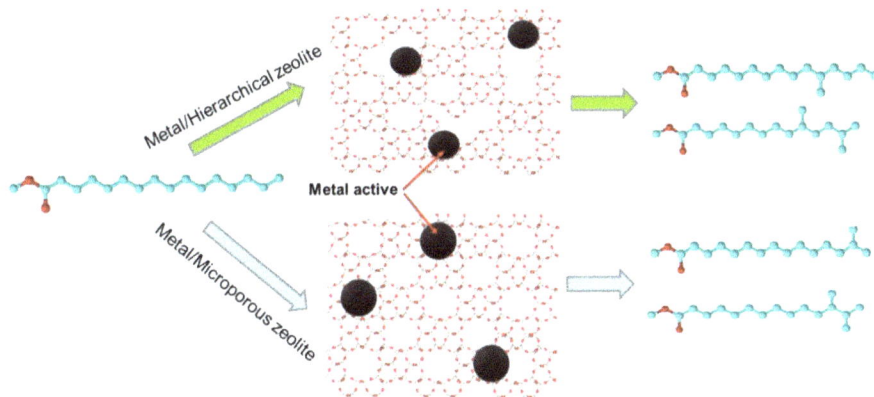

Figure 2. Schematic illustration of the effect of a hierarchical mesoporous ZSM-22 zeolite-based catalyst on hydroisomerization selectivity.

Table 6. Freezing points of some normal and isomethyl paraffins of various carbon numbers (modified from Reference [57,58]).

Carbon Number	Freezing Point, °C		
	n-Paraffin	2-Methylparaffin	5-Methylparaffin
C12	−10	−46	−70
C13	−5	−26	−69
C16	18	−10	−31
C18	28	6	−20
C20	37	18	−7

For *n*-decane, initially, both conventional and hierarchical ZSM-22 favor the formation of 2-methylnonane and hinder the formation of 4- and 5-methylnonanes. As the isomer yields approach their maximums, the positional distribution of methylnonanes reaches an equilibrium composition.

The proportion of 2- over 5-methyl-nonane generation at 5% conversion is termed the refined constraint index CI°, reflecting the pore width [59]. After hierarchization, the CI° drops from 14.5 to 8.0, which is still in the range of the 10-R zeolite (CI° > 2.2) [59]. The efficient catalytic hydroisomerization observed for the hierarchized catalysts can be explained by the rearrangement of the spatial Brønsted acidity distribution. Namely, hierarchization reduces the number of acid sites in the micropores, thereby limiting the hydrocracking reaction. At the same time, it increases the number of acid sites at pore mouths, thereby favoring the isomerization reaction. This argument is also investigated in research by Tao et al. [32], which focused on the preparation of hierarchical SAPO-11 zeolite by means of a dry-gel conversion method with 3-(trimethoxysilyl) propyl]octadecyldimethyl ammonium chloride (TPOAC) templating and its application for n-dodecane hydroisomerization. Besides the integration of mesopores into microporous zeolite, acidity tuning was also carried out to improve the catalysis performance of hierarchical zeolite. It is mostly believed that the medium and strong acidity of Brønsted acid sites plays a crucial role in skeletal isomerization. Notably, although the number of Brønsted acid sites of hierarchical SAPO-11 is less than that of conventional SAPO-11, the hydroisomerization conversion over Pt/hierarchical SAPO-11 is moderately higher than that over its nonhierarchical counterpart. This is likely because of the greater availability of pore mouths in the case of the hierarchical SAPO-11 zeolite-based catalyst. Similarly, the uniform intercrystalline mesopores of the hierarchical structure and the increased number of medium-acidity sites present owing to water content control can enhance the diffusion of the multibranched isomer out of the micropores before cracking occurs, thereby leading to higher yields of multibranched isomers.

2.2.3. Metal/Acid Site Balance

The interactivity between metal and zeolite plays an important role in bifunctional hydroisomerization catalysts. Mendes et al. [31] investigated the role of hydrogenating (metal)/acid function balance upon isomerized selectivity, representing this balance by the ratio between the number of accessible metal atoms (n_{Pt}) and the number of aluminum atoms (n_{Al}) that form the negative charge (AlO_4^-) in the zeolite's framework. Platinum-based catalysts supported on HUSY and HBEA zeolites showed increasing yields of iso-C_{16} at low n_{Pt}/n_{Al} values, followed by a plateau in yield for n_{Pt}/n_{Al} values higher than 0.01 and 0.02, respectively. The balance between dehydrogenating sites and acid sites is indicated by the maximum yield of iso-C_{16}. Increasing the metal loading (Table 7) causes dehydrogenation to take place much faster than acid-catalyzed steps, and thus the isoalkenes formed will be promptly hydrogenated, favoring isomerization rather than cracking.

Table 7. Metal dispersions, number of accessible metal atoms, metal/acid site ratios, corresponding turnover frequency (TOF_{Al}) at 480 K, and maximal isomer yields estimated acid site concentrations for selected catalysts with various Pt loadings (modified from Reference [31]).

Catalyst	n_{Al} (μmol/g)	Dispersion (%)	n_{Pt} (μmol/g)	n_{Pt}/n_{Al}	TOF_{Al} at 480 K (10^3 s^{-1})	Maximal i-C_{16} yield (%)
0.1%Pt/HUSY	820	53	3	0.003	1.6	46.5
0.4%Pt/HUSY	820	53	11	0.013	2.1	59.3
0.7%Pt/HUSY	820	52	19	0.023	1.6	62.1
0.1%Pt/HBEA	840	45	2	0.003	3.0	29.2
0.4%Pt/HBEA	840	43	9	0.010	4.0	63.5
1%Pt/HBEA	840	37	19	0.023	4.7	67.3

Similar dependences were also reported by Batalha et al. [60], who demonstrated that at low C_{Pt}/C_{H+} values the rate-determining step of n-C_{16} hydroisomerization over Pt/alumina-HBEA is the hydrogenation/dehydrogenation step over the metallic phase, and at high C_{Pt}/C_{H+} values it is the rearrangement of cycloprotonated intermediates over the acidic sites. On the Pd/SAPO-11-based catalyst [30], the optimal C_{Pd}/C_{H+} ratio is in the range of 0.09–0.27, corresponding to 0.1–0.3 wt % Pd loadings. This catalyst can be considered as an ideal hydroisomerization catalyst, showing excellent synergy between the metal Pd sites and Brønsted acid sites. The highest isomer selectivity of 89.9% at the conversion level of 90.8% was observed in *n*-decane transformation over the 0.3%Pd/SAPO-11 catalyst. For C_{Pd}/C_{H+} ratios higher than 0.47, the increasing agglomeration of Pd particles blocks the pore channel and pore mouth of the SAPO-11 structure, resulting in suppressed diffusion of the branched isomers and excessive cracking.

3. Blending Process

The blending of alternative fluids has been investigated as a means to improve the properties of vegetable oil-based insulating oils [61–64]. Usman et al. [61] studied several characteristics of a mixture of soybean and palm kernel oil for its suitability as an insulator fluid in power transformers. Although such blends have very high flash points (>234 °C) and economic and environmental advantages, these blended oils do not show any synergy in improving physical properties such as pour point and viscosity, which are dependent upon the degree of saturation. A study by Bertrand et al. [62] demonstrated the ability to decrease the pour point and viscosity of vegetable oil-based insulating oil by mixing oleic rapeseed oil and fatty monoesters in a 1:1 ratio and adding 0.3% of the inhibitor di-tert-butyl-para-cresol. The resulting blend showed the pour point of −30 °C and the kinetic viscosity at 40 °C of 17 mm^2/s.

Using an additive to reduce the pour points and to increase the oxidation stability of natural esters is also an efficient method, and has been applied in electric industrial processing. To improve the low-temperature properties, polymer additives called pour point depressants (PPDs) can be used to reduce the pour point. There is adequate literature on the efficacy of various PPDs based on polymers such as modified carboxyl polymers, acrylate polymers, nitrogen-based acrylate polymers, and methylene-linked aromatic components. Unfortunately, the use of such PPDs can be limited by their nonbiodegradability. Antioxidant additives that have been applied to vegetable oil-based insulating oil include phenolic compounds such as butylated hydroxyanisole, butylated hydroxytoluene, tert-butylhydroquinone, and propyl gallate. Their role is to slow natural oxidation by reacting with free radicals to form stable compounds that do not rapidly react with oxygen. Many works have demonstrated the effects of such additives upon the oxidation stability of insulators [48,65–68]. Nonetheless, the economic costs and the human health and environmental impacts of these additives hinder their application [69,70].

4. Conclusions

Vegetable oils can be used as the feedstock of insulating oils for electric transformers thanks to their environmental benefits such as low ecotoxicity, high biodegradability, and replacement of petroleum oils. The main challenges in applying vegetable oils as insulating transformer oils are their poor low-temperature properties and low oxidation stabilities; owing to their saturated and unsaturated fatty acid esters, vegetable oils' pour points and viscosities are higher than those of mineral oils.

The hydroisomerization of saturated and unsaturated long-chain normal paraffins to form branched isomers is a promising pathway for upgrading vegetable oil-based insulating oils, even for the bio-based lubricant with a similar composition. The processing efficiency is determined by the catalyst used. Zeolites are of interest as catalyst supports for isomerization upgrading because of their microporous channels and lack of large voids, making them more selective for the hydroisomerization of long-chain paraffins. Furthermore, the diffusion limitations of zeolites can be suppressed by means of methods that form hierarchical mesoporous/microporous structures.

Although many reports have advanced our knowledge on the role of the metal phase and Brønsted acid sites, as well as the effects of the pore structures of many kind of zeolites upon activity and selectivity, a more detailed understanding of the metal/acid functional balance of noble metal/zeolite catalysts is needed to widen their application and decrease their cost.

Acknowledgments: This research was supported by the Basic Science Research Program (2017R1A2B4007648) and the C1 Gas Refinery Program (2015M3D3A1A01064882) through the National Research Foundation of Korea (NRF), funded by the Ministry of Science and ICT.

Author Contributions: Dieu-Phuong Phan prepared a draft of the manuscript. Eun Yeol Lee coordinated the study and finalized the manuscript. All authors read and approved the manuscript.

Conflicts of Interest: The authors declare no conflicts of interest.

References

1. Sierota, A.; Rungis, J. Electrical insulating oils. I. Characterization and pre-treatment of new transformer oils. *IEEE Electr. Insul. Mag.* **1995**, *11*, 8–20. [CrossRef]
2. Rouse, T.O. Mineral insulating oil in transformers. *IEEE Electr. Insul. Mag.* **1998**, *14*, 6–16. [CrossRef]
3. Miller, R.E. Silicone Transformer Liquid: Use, Maintenance, and Safety. *IEEE Trans. Ind. Appl.* **1981**, *IA-17*, 463–468. [CrossRef]
4. Dolata, B.; Borsi, H.; Gockenbach, E. New synthetic ester fluid for the insulation of liquid immersed transformers. In Proceedings of the Conference Record of the 2006 IEEE International Symposium on Electrical Insulation, Toronto, ON, Canada, 11–14 June 2006; pp. 534–537.
5. Nagashree, A.N.; Champa, V.; Sumangala, B.V.; Nagabhushana, G.R. Suitability of natural vegetable seed oil as liquid dielectric coolant in an insulation system. In Proceedings of the 2015 International Conference on Emerging Research in Electronics, Computer Science and Technology (ICERECT), Mandya, India, 17–19 December 2015; pp. 429–434.
6. McShane, C.P. Relative properties of the new combustion-resist vegetable-oil-based dielectric coolants for distribution and power transformers. *IEEE Trans. Ind. Appl.* **2001**, *37*, 1132–1139. [CrossRef]
7. Fernández, I.; Ortiz, A.; Delgado, F.; Renedo, C.; Pérez, S. Comparative evaluation of alternative fluids for power transformers. *Electr. Power Syst. Res.* **2013**, *98*, 58–69. [CrossRef]
8. Ramos, M.J.; María Fernández, C.; Casas, A.; Rodriguez, L.; Pérez, Á. Influence of fatty acid composition of raw materials on biodiesel properties. *Bioresour. Technol.* **2008**, *100*, 261–268. [CrossRef] [PubMed]
9. Al-Eshaikh, M.A.; Qureshi, M.I. Evaluation of Food Grade Corn Oil for Electrical Applications. *Int. J. Green Energy* **2012**, *9*, 441–455. [CrossRef]
10. Rafiq, M.; Lv, Y.Z.; Zhou, Y.; Ma, K.B.; Wang, W.; Li, C.R.; Wang, Q. Use of vegetable oils as transformer oils—A review. *Renew. Sustain. Energy Rev.* **2015**, *52*, 308–324. [CrossRef]
11. Xu, Y.; Qian, S.; Liu, Q.; Wang, Z.D. Oxidation stability assessment of a vegetable transformer oil under thermal aging. *IEEE Trans. Dielectr. Electr. Insul.* **2014**, *21*, 683–692. [CrossRef]
12. Bertrand, Y.; Hoang, L.C. Vegetal oils as substitute for mineral oils. In Proceedings of the 7th International Conference on Properties and Applications of Dielectric Materials (Cat. No. 03CH37417), Nagoya, Japan, 1–5 June 2003; Volume 492, pp. 491–494.
13. Jie, D.; Wang, Z.D.; Dyer, P.; Darwin, A.W.; James, I. Investigation of the impregnation of cellulosic insulations by ester fluids. In Proceedings of the 2007 Annual Report-Conference on Electrical Insulation and Dielectric Phenomena, Vancouver, BC, Canada, 14–17 October 2007; pp. 588–591.
14. Ngo, H.L.; Nuñez, A.; Lin, W.; Foglia, T.A. Zeolite-catalyzed isomerization of oleic acid to branched-chain isomers. *Eur. J. Lipid Sci. Technol.* **2007**, *109*, 214–224. [CrossRef]
15. Abhari, R.; Roth, E.G.; Havlik, P.Z.; Tomhnson, H.L. Bio-Based Synthetic Fluids. U.S. Patent 8,969,259 B2, 3 March 2015.
16. Reaume, S.; Ellis, N. Use of Isomerization and Hydroisomerization Reactions to Improve the Cold Flow Properties of Vegetable Oil Based Biodiesel. *Energies* **2013**, *6*, 619. [CrossRef]
17. Reaume, S.J.; Ellis, N. Synergistic Effects of Skeletal Isomerization on Oleic and Palmitic Acid Mixtures for the Reduction in Cloud Point of Their Methyl Esters. *Energy Fuels* **2012**, *26*, 4514–4520. [CrossRef]
18. Weitkamp, J. Isomerization of long-chain n-alkanes on a Pt/CaY zeolite catalyst. *Ind. Eng. Chem. Prod. Res. Dev.* **1982**, *21*, 550–558. [CrossRef]
19. Höchtl, M.; Jentys, A.; Vinek, H. Alkane conversion over Pd/SAPO molecular sieves: Influence of acidity, metal concentration and structure. *Catal. Today* **2001**, *65*, 171–177. [CrossRef]

20. Choudhury, I.R.; Hayasaka, K.; Thybaut, J.W.; Laxmi Narasimhan, C.S.; Denayer, J.F.; Martens, J.A.; Marin, G.B. Pt/H-ZSM-22 hydroisomerization catalysts optimization guided by Single-Event MicroKinetic modeling. *J. Catal.* **2012**, *290*, 165–176. [CrossRef]

21. Zhang, S.; Zhang, Y.; Tierney, J.W.; Wender, I. Anion-modified zirconia: Effect of metal promotion and hydrogen reduction on hydroisomerization of n-hexadecane and Fischer–Tropsch waxes. *Fuel Process. Technol.* **2001**, *69*, 59–71. [CrossRef]

22. De Lucas, A.; sánchez, P.; Dorado, F.; Ramos, M.J.; Valverde, J. Effect of the metal loading in the hydroisomerization of n-octane over beta agglomerated zeolite based catalysts. *Appl. Catal. A Gen.* **2005**, *294*, 215–225. [CrossRef]

23. Akhmedov, V.M.; Al-Khowaiter, S.H. Recent Advances and Future Aspects in the Selective Isomerization of High n-Alkanes. *Catal. Rev.* **2007**, *49*, 33–139. [CrossRef]

24. Martens, J.A.; Verboekend, D.; Thomas, K.; Vanbutsele, G.; Gilson, J.P.; Perez-Ramirez, J. Hydroisomerization of emerging renewable hydrocarbons using hierarchical Pt/H-ZSM-22 catalyst. *ChemSusChem* **2013**, *6*, 421–425. [CrossRef] [PubMed]

25. Yang, J.; Kikhtyanin, O.V.; Wu, W.; Zhou, Y.; Toktarev, A.V.; Echevsky, G.V.; Zhang, R. Influence of the template on the properties of SAPO-31 and performance of Pd-loaded catalysts for n-paraffin isomerization. *Microporous Mesoporous Mater.* **2012**, *150*, 14–24. [CrossRef]

26. Guisnet, M. "Ideal" bifunctional catalysis over Pt-acid zeolites. *Catal. Today* **2013**, *218*, 123–134. [CrossRef]

27. Coonradt, H.L.; Garwood, W.E. Mechanism of Hydrocracking. Reactions of Paraffins and Olefins. *Ind. Eng. Chem. Process Des. Dev.* **1964**, *3*, 38–45. [CrossRef]

28. Parmar, S.; Pant, K.K.; John, M.; Kumar, K.; Pai, S.M.; Newalkar, B.L. Hydroisomerization of Long Chain n-Paraffins over Pt/ZSM-22: Influence of Si/Al Ratio. *Energy Fuels* **2015**, *29*, 1066–1075. [CrossRef]

29. Niu, P.; Xi, H.; Ren, J.; Lin, M.; Wang, Q.; Jia, L.; Hou, B.; Li, D. High selectivity for n-dodecane hydroisomerization over highly siliceous ZSM-22 with low Pt loading. *Catal. Sci. Technol.* **2017**, *7*, 5055–5068. [CrossRef]

30. Song, X.; Bai, X.; Wu, W.; Kikhtyanin, O.V.; Zhao, A.; Xiao, L.; Su, X.; Zhang, J.; Wei, X. The effect of palladium loading on the catalytic performance of Pd/SAPO-11 for n-decane hydroisomerization. *Mol. Catal.* **2017**, *433*, 84–90. [CrossRef]

31. Mendes, P.S.F.; Mota, F.M.; Silva, J.M.; Ribeiro, M.F.; Daudin, A.; Bouchy, C. A systematic study on mixtures of Pt/zeolite as hydroisomerization catalysts. *Catal. Sci. Technol.* **2017**, *7*, 1095–1107. [CrossRef]

32. Tao, S.; Li, X.L.; Lv, G.; Wang, C.X.; Xu, R.S.; Ma, H.J.; Tian, Z.J. Highly mesoporous SAPO-11 molecular sieves with tunable acidity: Facile synthesis, formation mechanism and catalytic performance in hydroisomerization of n-dodecane. *Catal. Sci. Technol.* **2017**, *7*, 5775–5784. [CrossRef]

33. Snåre, M.; Kubičková, I.; Mäki-Arvela, P.; Eränen, K.; Murzin, D.Y. Heterogeneous Catalytic Deoxygenation of Stearic Acid for Production of Biodiesel. *Ind. Eng. Chem. Res.* **2006**, *45*, 5708–5715. [CrossRef]

34. Bauer, F.; Karsten, F.; Marko, B.; Wolf-Dietrich, E.; Thomas, K.; Roger, G. Hydroisomerization of Long-Chain Paraffins over Nano-Sized Bimetallic Pt-Pd/H-Beta Catalysts. *Catal. Sci. Technol.* **2014**, *4*, 4045–4054. [CrossRef]

35. Blomsma, E.; Martens, J.A.; Jacobs, P.A. Isomerization and Hydrocracking of Heptane over Bimetallic Bifunctional PtPd/H-Beta and PtPd/USY Zeolite Catalysts. *J. Catal.* **1997**, *165*, 241–248. [CrossRef]

36. Sachtler, M.H.; Zhang, Z. Zeolite-Supported Transition Metal Catalysts. *Adv. Catal.* **1993**, *39*, 129–220.

37. Lee, J.-K.; Rhee, H.-K. Sulfur tolerance of zeolite beta-supported Pd−Pt catalysts for the isomerization of n-hexane. *J. Catal.* **1998**, *177*, 208–216. [CrossRef]

38. Pawelec, B.; Mariscal, R.; Navarro, R.M.; van Bokhorst, S.; Rojas, S.; Fierro, J.L.G. Hydrogenation of aromatics over supported Pt-Pd catalysts. *Appl. Catal. A Gen.* **2002**, *225*, 223–237. [CrossRef]

39. Verma, D.; Kumar, R.; Rana, B.S.; Sinha, A.K. Aviation fuel production from lipids by a single-step route using hierarchical mesoporous zeolites. *Energy Environ. Sci.* **2011**, *4*, 1667–1671. [CrossRef]

40. Mériaudeau, P.; Tuan, V.A.; Nghiem, V.T.; Lai, S.Y.; Hung, L.N.; Naccache, C. SAPO-11, SAPO-31, and SAPO-41 Molecular Sieves: Synthesis, Characterization, and Catalytic Properties inn-Octane Hydroisomerization. *J. Catal.* **1997**, *169*, 55–66. [CrossRef]

41. Li, S.; Zheng, A.; Su, Y.; Zhang, H.; Chen, L.; Yang, J.; Ye, C.; Deng, F. Brønsted/Lewis Acid Synergy in Dealuminated HY Zeolite: A Combined Solid-State NMR and Theoretical Calculation Study. *J. Am. Chem. Soc.* **2007**, *129*, 11161–11171. [CrossRef] [PubMed]

42. Maesen, T.L.M.; Schenk, M.; Vlugt, T.J.H.; Jonge, J.P.d.; Smit, B. The Shape Selectivity of Paraffin Hydroconversion on TON-, MTT-, and AEL-Type Sieves. *J. Catal.* **1999**, *188*, 403–412. [CrossRef]
43. Claude, M.C.; Martens, J.A. Monomethyl-Branching of Long n-Alkanes in the Range from Decane to Tetracosane on Pt/H-ZSM-22 Bifunctional Catalyst. *J. Catal.* **2000**, *190*, 39–48. [CrossRef]
44. Weisz, P.B.; Frilette, V.J.; Maatman, R.W.; Mower, E.B. Catalysis by crystalline aluminosilicates II. Molecular-shape selective reactions. *J. Catal.* **1962**, *1*, 307–312. [CrossRef]
45. Martens, J.A.; Verboekend, D.; Thomas, K.; Vanbutsele, G.; Pérez-Ramírez, J.; Gilson, J.-P. Hydroisomerization and hydrocracking of linear and multibranched long model alkanes on hierarchical Pt/ZSM-22 zeolite. *Catal. Today* **2013**, *218–219*, 135–142. [CrossRef]
46. Martens, J.A.; Souverijns, W.; Verrelst, W.; Parton, R.F.; Froment, G.; Pierre, J. Selective Isomerization of Hydrocarbon Chains on External Surfaces of Zeolite Crystals. *Angew. Chem. Int. Ed.* **1995**, *34*, 2528–2530. [CrossRef]
47. Wei, Y.; Parmentier, T.E.; de Jong, K.P.; Zecevic, J. Tailoring and visualizing the pore architecture of hierarchical zeolites. *Chem. Soc. Rev.* **2015**, *44*, 7234–7261. [CrossRef] [PubMed]
48. Ullah, J.; Hamayoun, M.; Ahmad, T.; Ayub, M.; Zafarullah, M. Effect of light, natural and synthetic antioxidants on stability of edible oil and fats. *Asian J. Plant Sci.* **2003**, *2*, 1192–1194.
49. Wang, Y.; Tao, Z.; Wu, B.; Xu, J.; Huo, C.; Li, K.; Chen, H.; Yang, Y.; Li, Y. Effect of metal precursors on the performance of Pt/ZSM-22 catalysts for n-hexadecane hydroisomerization. *J. Catal.* **2015**, *322*, 1–13. [CrossRef]
50. Shi, Y.; Cao, Y.; Duan, Y.; Chen, H.; Chen, Y.; Yang, M.; Wu, Y. Upgrading of palmitic acid to iso-alkanes over bi-functional Mo/ZSM-22 catalysts. *Green Chem.* **2016**, *18*, 4633–4648. [CrossRef]
51. Claude, M.C.; Vanbutsele, G.; Martens, J.A. Dimethyl Branching of Long n-Alkanes in the Range from Decane to Tetracosane on Pt/H–ZSM-22 Bifunctional Catalyst. *J. Catal.* **2001**, *203*, 213–231. [CrossRef]
52. Nghiem, V.T.; Sapaly, G.; Mériaudeau, P.; Naccache, C. Monodimensional tubular medium pore molecular sieves for selective hydroisomerisation of long chain alkanes: n-octane reaction on ZSM and SAPO type catalysts. *Top. Catal.* **2000**, *14*, 131–138. [CrossRef]
53. Chi, K.B.; Zhao, Z.; Tian, Z.J.; Hu, S.; Yan, L.J.; Li, T.S.; Wang, B.C.; Meng, X.B.; Gao, S.B.; Tan, M.W.; et al. Hydroisomerization performance of platinum supported on ZSM-22/ZSM-23 intergrowth zeolite catalyst. *Pet. Sci.* **2013**, *10*, 242–250. [CrossRef]
54. Vandegehuchte, B.D.; Thybaut, J.W.; Martens, J.A.; Marin, G.B. Maximizing n-alkane hydroisomerization: The interplay of phase, feed complexity and zeolite catalyst mixing. *Catal. Sci. Technol.* **2015**, *5*, 2053–2058. [CrossRef]
55. Liu, Y.; Qu, W.; Chang, W.; Pan, S.; Tian, Z.; Meng, X.; Rigutto, M.; Made, A.v.d.; Zhao, L.; Zheng, X.; et al. Catalytically active and hierarchically porous SAPO-11 zeolite synthesized in the presence of polyhexamethylene biguanidine. *J. Colloid Interface Sci.* **2014**, *418*, 193–199. [CrossRef] [PubMed]
56. Liu, Z.; Liu, L.; Song, H.; Wang, C.; Xing, W.; Komarneni, S.; Yan, Z. Hierarchical SAPO-11 preparation in the presence of glucose. *Mater. Lett.* **2015**, *154*, 116–119. [CrossRef]
57. Krár, M.; Kasza, T.; Kovács, S.; Kalló, D.; Hancsók, J. Bio gas oils with improved low temperature properties. *Fuel Process. Technol.* **2011**, *92*, 886–892. [CrossRef]
58. Knothe, G.; Dunn, R. A Comprehensive Evaluation of the Melting Points of Fatty Acids and Esters Determined by Differential Scanning Calorimetry. *J. Am.Oil Chem. Soc.* **2009**, *86*, 843–856. [CrossRef]
59. Martens, J.A.; Jacobs, P.A. The potential and limitations of the n-decane hydroconversion as a test reaction for characterization of the void space of molecular sieve zeolites. *Zeolites* **1986**, *6*, 334–348. [CrossRef]
60. Batalha, N.; Pinard, L.; Bouchy, C.; Guillon, E.; Guisnet, M. n-Hexadecane hydroisomerization over Pt-HBEA catalysts. Quantification and effect of the intimacy between metal and protonic sites. *J. Catal.* **2013**, *307*, 122–131. [CrossRef]
61. Usman, M.A.; Olanipekun, O.O.; Henshaw, U.T. A Comparative Study of Soya Bean Oil and Palm Kernel Oil as Alternatives to Transformer Oil. *J. Emerg. Trends Eng. Appl. Sci.* **2012**, *3*, 33–37.
62. Bertrand, Y.; Lauzevis, P. development of a low viscosity insulating liquid based on natural esters for distribution transformers. In Proceedings of the 22nd International Conference on Electricity Distribution, Stockholm, Sweden, 10–13 June 2013; p. 382.

63. Suwarno; Darma, I.S. Dielectric properties of mixtures between mineral oil and natural ester. In Proceedings of the 2008 International Symposium on Electrical Insulating Materials (ISEIM 2008), Mie, Japan, 7–11 September 2008; pp. 514–517.

64. Qiu, H.B.S.W. Blended Oil Compositions Useful as Dielectric Fluid Compositions and Methods of Preparing Same. U.S. Patent US20140131634A1, 13 November 2012.

65. Dunn, R.O. Effect of antioxidants on the oxidative stability of methyl soyate (biodiesel). *Fuel Process. Technol.* **2005**, *86*, 1071–1085. [CrossRef]

66. Jung, M.Y.; Min, D.B. Effects of oxidized α-, γ- and δ-tocopherols on the oxidative stability of purified soybean oil. *Food Chem.* **1992**, *45*, 183–187. [CrossRef]

67. Ruger, C.W.; Klinker, E.J.; Hammond, E.G. Abilities of some antioxidants to stabilize soybean oil in industrial use conditions. *J. Am. Oil Chem. Soc.* **2002**, *79*, 733–736. [CrossRef]

68. Kumar, S.S.; Iruthayarajan, M.W.; Bakrutheen, M.; Kannan, S.G. Effect of antioxidants on critical properties of natural esters for liquid insulations. *IEEE Trans. Dielectr. Electr. Insul.* **2016**, *23*, 2068–2078. [CrossRef]

69. Khanahmadi, M.; Janfeshan, K. Study on Antioxidation Property of Ferulago angulata Plant. *Asian J. Plant Sci.* **2006**, *5*, 521–526.

70. Morteza-Semnani, K.; Saeedi, M.; Shahani, S. Antioxidant activity of the methanolic extracts of some species of Phlomis and Stachys on sunflower oil. *Afr. J. Biotechnol.* **2006**, *5*, 2428–2432.

catalysts

MDPI

Article

Desilicated ZSM-5 Zeolites for the Production of Renewable *p*-Xylene via Diels–Alder Cycloaddition of Dimethylfuran and Ethylene

Joel McGlone, Peter Priecel, Luigi Da Già, Liqaa Majdal and Jose A. Lopez-Sanchez *

Stephenson Institute for Renewable Energy, Department of Chemistry, University of Liverpool, Liverpool L69 7ZD, UK; jomcgl1@hotmail.com (J.M.); peter.priecel@liverpool.ac.uk (P.P.); davia.luigi@gmail.com (L.D.V.); L.I.Majdal@liverpool.ac.uk (L.M.)
* Correspondence: jals@liverpool.ac.uk; Tel.: +44-151-79-43535

Received: 30 April 2018; Accepted: 14 June 2018; Published: 20 June 2018

Abstract: The selective production of *p*-xylene and other aromatics starting from sugars and bioderived ethylene offers great promise and can eliminate the need for separation of xylene isomers, as well as decreasing dependency on fossil resources and CO_2 emissions. Although the reaction is known, the microporosity of traditional commercial zeolites appears to be a limiting factor. In this work, we demonstrate for the first time that simply desilication of microporous commercial zeolites by a simple NaOH treatment can greatly enhance conversion and selectivity. The [4 + 2] Diels–Alder cycloaddition of 2,5-dimethylfuran with ethylene in a pressurised reactor was investigated using a series of H-ZSM-5 catalysts with SiO_2/Al_2O_3 ratios 30 and 80 with increasing pore size induced by desilication. X-ray diffraction, scanning electron microscopy, ^{27}Al magic-angle spinning nuclear magnetic resonance, temperature programmed desorption of ammonia, and nitrogen physisorption measurements were used to characterise the catalysts. The enhancement of conversion was observed for all desilicated samples compared to the untreated zeolite, and increases in temperature and ethylene pressure significantly improved both dimethylfuran conversion and selectivity to *p*-xylene due to the easier desorption from the zeolite's surface and the augmented cycloaddition rate, respectively. A compromise between acidity and mesoporosity was found to be the key to enhancing the activity and maximising the selectivity in the production of *p*-xylene from 2,5-dimethylfuran.

Keywords: renewable *p*-xylene; Diels–Alder; desilication; dimethylfuran; biomass; ZSM-5; hierarchical zeolites; renewable aromatics

1. Introduction

Due to diminishing quantities of fossil fuels, lignocellulosic biomass is attracting more attention as a feedstock for platform chemicals due to its large availability and economic factors [1]. One of these platform chemicals obtainable from biomass is *p*-xylene [2,3], a precursor in the terephthalic acid synthesis, which can be polymerised to produce polyethylene terephthalate (PET). *p*-Xylene is currently produced on an industrial scale from the catalytic reforming of petroleum naphtha, which is part of the benzene-toluene-xylene (BTX) fraction. Current market demand for *p*-xylene is typically around 34% of the BTX fraction [4].

Brion et al. [5] found that furan ring compounds could be converted into a six-membered ring via cycloaddition to produce an oxa-norbornene intermediate followed by β elimination (Scheme 1). Recent studies have shown that 2,5-dimethylfuran (DMF) could act as a potential feedstock for the production of *p*-xylene [6–10]. The advantage of this route is the production of the pure para isomer of xylene, thus removing the necessity of purifying *p*-xylene, as is the case in its production from fossil sources. DMF itself can be produced from glucose in a two-step process. Glucose can be converted into 2,5-hydroxymethylfurfural (HMF), which can then be hydrodeoxygenated to produce DMF (Scheme 1) [11].

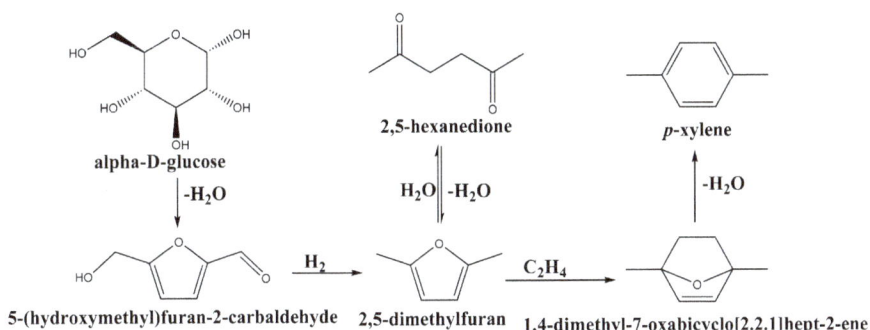

Scheme 1. Proposed reaction for the conversion of glucose to *p*-xylene.

The production of *p*-xylene was reported via the Diels–Alder cycloaddition of DMF and ethylene over different acid catalysts, such as zeolite FAU (faujasite) [6,12,13], γ-alumina, niobic acid, H-FAU (ZSM-5) [14], or zeolite BEA (beta) [9,15,16]. H-BEA was shown to be almost an order of magnitude more active than γ-alumina, niobic acid, or H-ZSM-5. It was recently discussed that both Brønsted and Lewis acids are very effective in catalysing Diels–Alder cycloaddition [3], and several teams demonstrated enhancement in catalytic activity of zeolites containing both Brønsted and Lewis acid sites [9,17,18]. Additionally, it was shown that polar aprotic solvent, such as 1,4-dioxane, improves both DMF conversion and *p*-xylene yield attributable to the increased dehydration rate. Several studies were dedicated to the detailed study of the reaction mechanism and regimes reflecting different rate-limiting steps and thermodynamics [6,7,12,13,16,19]. It was found that regimes of high and low acid concentration differ significantly and result in switching the rate-limiting step from the Diels–Alder cycloaddition to the dehydration of oxanorbornene, respectively.

It was reported that inducing mesoporosity in zeolites (producing hierarchical zeolites) can alleviate diffusion limitations and significantly increase catalytic activity in a variety of reactions [14,20]. It has been argued that the secondary porosity could also be used to store coke and therefore avoid blockage of the active microporous sites [21]. One of the ways to introduce mesoporosity in zeolites is the removal of framework Si atoms (desilication) [22]. Dean [23] first suggested the idea of alkaline treatment for zeolites to increase performance in the mordenite in the gas–oil reaction and noted that crystallinity was preserved with a three times higher conversion compared to the untreated parent zeolite. Groen et al. [22] found that desilication preserved crystallinity and Brønsted acidity of the parent zeolite.

In view of this, we decided to produce hierarchical zeolites for the Diels–Alder cycloaddition of ethylene to DMF with the hope that increasing pore size would facilitate mass transport phenomena and increase reaction rates. In this paper, we investigate the effect of increasing the pore size of ZSM-5 catalysts with differing SiO_2/Al_2O_3 ratios by alkaline treatment and how this affects the structural properties of the ZSM-5, as well as how beneficial this can be for the production of renewable aromatics *p*-xylene.

2. Results and Discussion

2.1. X-ray Diffraction

XRD patterns of the untreated and alkaline-treated samples were obtained and are shown in Figures 1 and 2 for Z30 and Z80, respectively. All samples show diffractions of ZSM-5 based on JCPDS

PDF 01-080-0922. The parent zeolite was compared to the alkaline-treated ones, and intensity of the (101) peak was used for the crystallinity calculation.

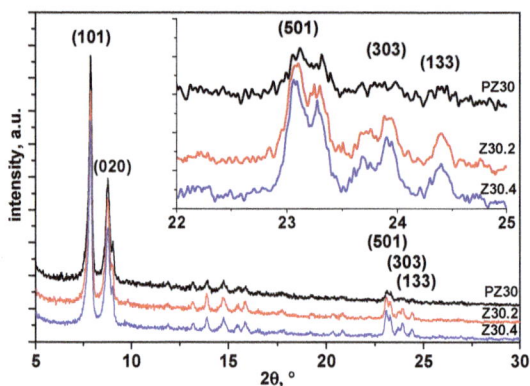

Figure 1. XRD patterns of PZ30 (top, black), Z30.2 (middle, red), and Z30.2 (bottom, blue).

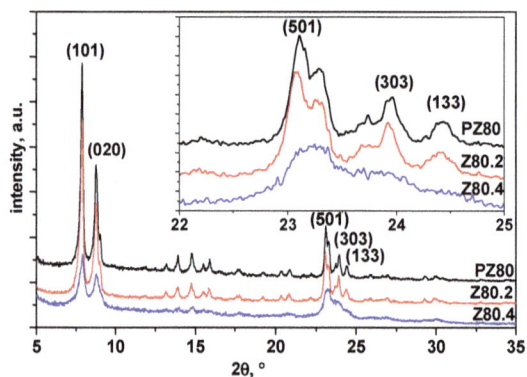

Figure 2. XRD patterns of PZ80 (top, black), Z80.2 (middle, red), and Z80.2 (bottom, blue).

It was found that 0.2 M NaOH treatments of PZ30 and PZ80 did not significantly reduce the material crystallinity. Alkaline treatment of Z30 (up to 0.4 M NaOH) preserved crystallinity up to 87% of the original zeolite, as evidenced by the presence of diffraction lines (101), (020), and (501), which are typical of ZSM-5 diffraction patterns (Figure 1).

On the other hand, the alkaline treatment of PZ80 with 0.4 M NaOH significantly reduced the zeolite crystallinity down to 31% of the original value (Figure 2). These results are in agreement with the hypothesis of Verboekend et al. [24], who claimed that desilication is more difficult at lower SiO_2/Al_2O_3 ratios due to the repulsion by the negative charge on the framework aluminium ions and would be easier with SiO_2/Al_2O_3 ratios >50 due to uncontrolled desilication.

2.2. Surface Area and Porosity of the Materials

N_2 isotherms of PZ30 and PZ80 display typical type I isotherm with a plateau at a higher relative pressure according to IUPAC (International Union of Pure and Applied Chemistry) classification (Figures 3 and 4).

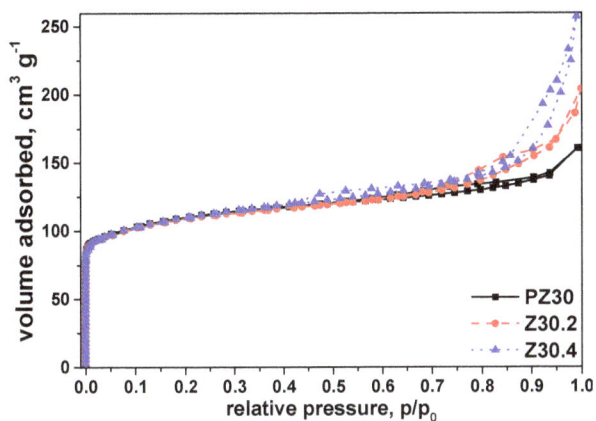

Figure 3. N$_2$ isotherms of PZ30 (black square, full line), Z30.2 (red circle, dashed line), and Z30.4 (blue triangle, dotted line) at 77 K.

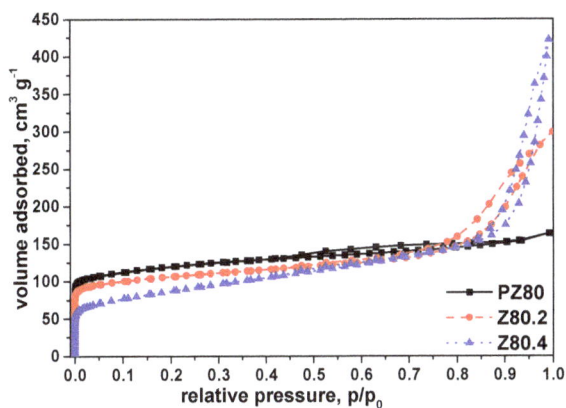

Figure 4. N$_2$ isotherms of PZ80 (black square, full line), Z80.2 (red circle, dashed line), and Z80.4 (blue triangle, dotted line) at 77 K.

In addition to this, both PZ30 and PZ80 exhibit a small hysteresis, which is indicative of limited mesoporosity. The shapes of the isotherms confirm that the structure of PZ30 and PZ80 is dominated by microporosity. N$_2$ isotherms of all alkaline-treated Z30 and Z80 show a combination of type I and type IV isotherms with an increased uptake of N$_2$ and pronounced hysteresis, which is indicative of increased mesoporosity. Brunauer–Emmett–Teller (BET) measurements showed that alkaline treatment had minimal effect on the surface area of Z30 (Table 1).

Table 1. Physical properties of untreated and treated H-ZSM-5 Z30 and Z80 catalysts.

Catalyst	S_{Bet} (m² g⁻¹) [a]	V_{total} [b] (cm³ g⁻¹)	V_{micro} [c] (cm³ g⁻¹)	S_{meso} [c] (m² g⁻¹)	Crystallinity (%) [d]	SiO₂/Al₂O₃ Ratio [e]	Total Acid Sites, μmol g⁻¹ [f]	Weak Acid Sites, μmol g⁻¹ [f]	Strong Acid Sites, μmol g⁻¹ [f]	Brønsted Acid Sites, μmol g⁻¹ [g]	Lewis Acid Sites, μmol g⁻¹ [g]
PZ30	344	0.250	0.147	142	100	28	581	165	416	295	286
Z30.2	340	0.315	0.147	181	92	24	655	205	450	163	492
Z30.4	340	0.397	0.148	216	87	21	692	266	426	337	355
PZ80	376	0.253	0.165	131	100	72	333	55	278	166	167
Z80.2	327	0.454	0.143	224	84	50	369	109	260	24	345
Z80.4	312	0.734	0.122	260	31	31	504	260	244	368	136

[a] Brunauer–Emmett–Teller (BET) method; [b] $P/P_0 = 0.99$; [c] NLDFT (non-local density functional theory); [d] measured by XRD; [e] ICP-OES (inductively coupled plasma-optical emission spectroscopy; [f] NH₃-TPD; [g] B/L (Brønsted/Lewis) acid site ratio from pyridine FTIR and to total amount from TPD (temperature programmed desorption).

Upon the treatment with 0.2 M and 0.4 M NaOH, the mesopore surface area of Z30 increased from 142 to 181 and 216 m^2 g^{-1} compared to the parent zeolite, while the micropore volume was preserved.

In the case of Z80, N$_2$ adsorption measurements showed that both the surface area and microporosity decreased after the NaOH treatment. The BET surface area decreased from 376 to 312 m^2 g^{-1} and the volume of micropores from 0.165 to 0.122 cm^3 g^{-1} for Z80.4 when compared to the untreated PZ80. The alkaline treatment also had a profound effect on the mesopore surface area, which increased from 131 (PZ80) to 260 m^2 g^{-1} (Z80.4). These porosity changes in both Z30 and Z80 further corroborate the difference in difficult desilication at lower SiO$_2$/Al$_2$O$_3$ ratios and uncontrolled desilication at SiO$_2$/Al$_2$O$_3$ 80 [24].

2.3. Acidity Masurements

There were two desorption peaks in ammonia TPD (temperature programmed desorption) profiles (Figures S1 and S2). The first desorption peak at ~225 °C corresponds to the weak (Lewis) acid sites, whereas the second desorption peak at ~385 °C corresponds to the strong (both Brønsted and Lewis) acid sites [25,26]. It was reported that the concentration of the weakly bound ammonia observed at the low temperature peak has no catalytic importance, and it was demonstrated that it can be decreased by extending the flushing time with the inert gas [27].

The total acidity values of both the parent PZ30 and PZ80 are in agreement with the data present in the literature [27,28]. Both Z30 and Z80 desilicated samples showed increases in the total acidity, which is in agreement with the decrease in the SiO$_2$/Al$_2$O$_3$ ratio after the alkaline treatment; although, these results contradict the findings of Rac et al. [28], who found a decrease in the total acidity of all desilicated samples when compared to their corresponding parent. This could be due to the fact that during the desilication, the intensities of both the low and high temperature peaks increase. Also, Rac and co-workers used calorimetry compared to the TPD used in this study.

In the case of the parent PZ30 with a low SiO$_2$/Al$_2$O$_3$ ratio, the rigid zeolite structure and close proximity of the framework Si and Al (Lowenstein's rule) prevents the facile desilication to some extent, and the change in acidity profile is the least significant. This is also documented by the decrease in the SiO$_2$/Al$_2$O$_3$ ratio from 28 (PZ30) to 24 (Z30.2) and 21 (Z30.4) and the low decrease in crystallinity (<15%). As shown in Table 1, the total acidity for the Z30.2 and Z30.4 increased from 581 to 655 and 692 µmol g^{-1}, respectively. The most pronounced change in the acidity profile was found to come from the weak acid sites, which could be due to the increased mesoporosity, as it was suggested by Rodriguez-Gonzalez et al. [27] that weakly bound NH$_3$ can physisorb onto itself within the pores. The number of strong acid sites was not affected significantly and increased slightly for treated samples by max. 8% of the parent value, which is in good agreement with literature [25].

2.4. Microscopy

SEM measurements showed PZ30 and PZ80 displayed well-defined morphologies of compact particle sizes ca. 250 nm. The alkaline treatment with sodium hydroxide had contrasting effects on the morphologies of PZ30 and PZ80 (Figure 5). The external surfaces remained largely intact upon mild treatment with sodium hydroxide. This suggests that silicon extraction did not largely take place on the external surface of the zeolite crystals. Similar conclusions were reported by Sadowaska et al. [29]. Alkaline treatment had minimal effect on the morphology of PZ30, whereas for PZ80, there was a clear change in the morphology of the crystal structure after the alkaline treatment, as can be seen from the formation of a less compact structure (more images in Supplementary Information Figures S3 and S4). In detail, the transition from PZ30 to Z30.2 did not show significant changes in the shape of the crystallites, nor separation of the smaller crystals. In the case of Z80, the division of large crystallites into smaller bulks is clearly visible.

Figure 5. SEM images of untreated and treated HZSM-5. PZ30 (**A**); Z30.2 (**B**); PZ80 (**C**); and Z80.2 (**D**).

2.5. ^{27}Al MAS NMR

^{27}Al MAS NMR (magic angle spinning nuclear magnetic resonance) measurements were performed to determine the coordination of aluminium in the presented zeolite samples before and after the desilication treatment (Figure 6). The NMR spectra show peaks centred at ca. 54, 45–35, and 0 ppm, which correspond to framework aluminium in tetrahedral coordination (Al_{Td}) and extra-framework penta-coordinated and octahedral aluminium (Al_{Oh}), respectively [30].

As expected, the majority of Al occupies tetrahedral framework positions (Table 2). It can be seen that in both cases of Z30 and Z80, the isotropic shift of the 54 ppm resonance is widened as the concentration of the desilication agent is increased, which indicates the change of T-O-T angles in the framework (T is either Al or Si) [31]. This effect was more pronounced in the case of Z80, which agreed with its easier desilication and more pronounced decrease in SiO_2/Al_2O_3 ratio and crystallinity.

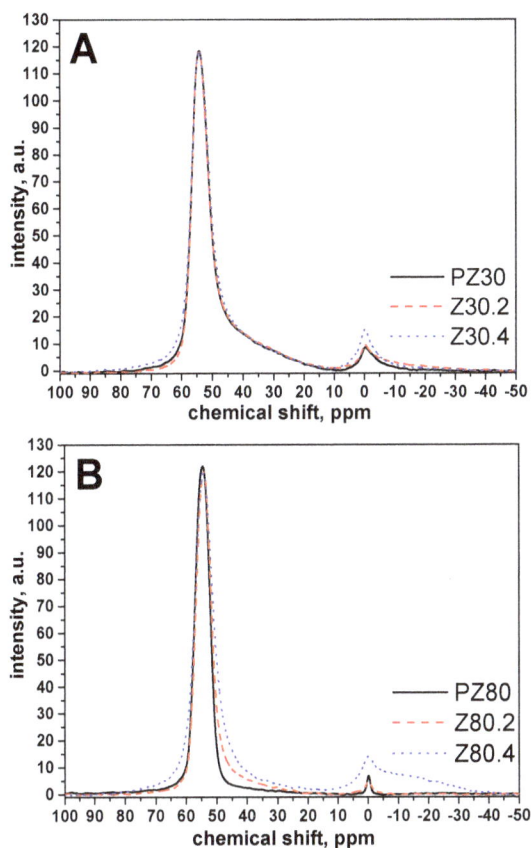

Figure 6. ^{27}Al MAS NMR spectra of presented zeolites with SiO_2/Al_2O_3 ratios 30 (Z30) (**A**) and 80 (Z80) (**B**) before (PZ30, PZ80) and after the desilication treatment (Z30.2, Z30.4, Z80.2, Z80.4).

Table 2. Normalised areas and ratios of peaks at 54 and 0 ppm corresponding to framework and extra-framework aluminium.

SAMPLE	Ratio of Peaks at 54 to 0 ppm	Normalised Areas at Peak Positions	
		54 ppm	0 ppm
PZ30	17.8	0.947	0.053
Z30.2	11.9	0.923	0.077
Z30.4	11.1	0.917	0.083
PZ80	63.7	0.985	0.015
Z80.2	42.5	0.977	0.023
Z80.4	4.7	0.823	0.177

A similar effect to that of framework aluminium was observed in the case of extra-framework Al in octahedral coordination. As the concentration of NaOH during the desilication increased, so did the amount of extra-framework aluminium. Moreover, the widening of the peak at ca. 0 ppm indicates a broader distribution of the local extra-framework Al environment, which was again more pronounced in the case of Z80.

Worth noting is the possible presence of penta-coordinated extra-framework aluminium, which could be suggested by the presence of the shoulder at 45–35 ppm. In our case, the relative amount of this type of Al does not seem to change during the treatment of Z30 but slowly increases along with the increment of Al_{Oh}.

2.6. Catalytic Activity

2.6.1. General Considerations and Conversion

As shown in Scheme 1, *p*-xylene is formed by the [4 + 2] cycloaddition of DMF with ethylene to form 7-oxabicyclo[2.2.1]hept-2-ene, which is dehydrated to form *p*-xylene. DMF can be also hydrolysed to form 2,5-hexanedione (HDO), which was shown to be able to dehydrate back to DMF and, according to this mechanism, could still be converted to *p*-xylene [12]. Trace amounts of 3-methyl-2-cyclopentenone (MCP) were found due to the intramolecular aldol reaction of 2,5-hexanedione. 1-ethyl-2,5-dimethylbenzene, which was found to be a product in the literature [6], was not found in our study. The reaction in which no catalyst was used yielded less than 3% conversion of DMF with no selectivity to any of the identified products. A typical time-dependent profile of the conversion of DMF and selectivities to the identified products is presented in Figure 7.

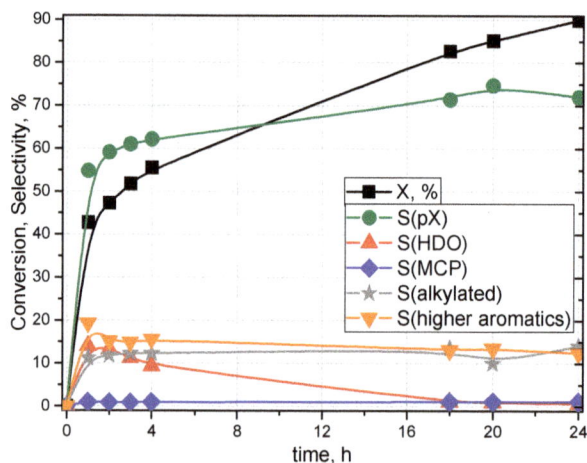

Figure 7. Conversion and selectivities to known products as a function of time for the Diels–Alder cycloaddition of 2,5-dimethylfuran (DMF) and ethylene over H-ZSM-5 (30). Reaction conditions: temperature, 250 °C; pressure, 55 bar; catalyst, 120 mg; 30 mL of 3 M DMF in *n*-octane; and stirring, 1100 rpm. Mass balance increased from 66% at first sampling to 73% at the end of the reaction.

As conversion of 2,5-dimethylfuran increases, *p*-xylene is produced, and its selectivity increase follows conversion until the last point, when it decreases slightly in favour of alkylated products. The selectivities to alkylated products, consisting of DMF and aromatics and higher aromatics with multiple aromatic rings, show an initial increase to ca. 10–15% after which they keep slightly decreasing until the end of the reaction. As mentioned before, hexanedione is produced but is again consumed, which suggests conversion back to DMF to produce *p*-xylene.

Conversion for both Z30 and Z80 treated by 0.2 M NaOH increased from 19.3% and 21.3% to 36% and 37%, respectively (Figures 8 and 9, Table S1).

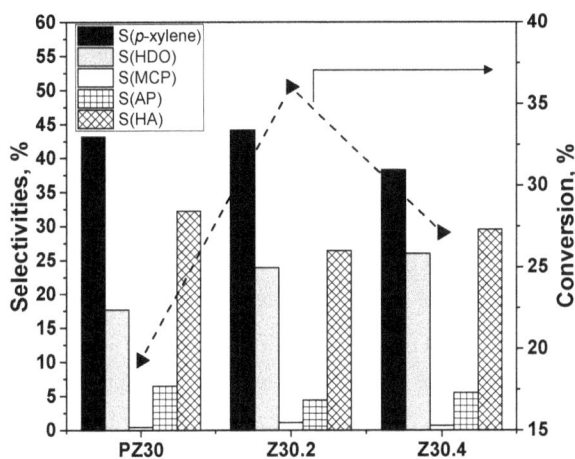

Figure 8. Conversion (X, symbol + line) and selectivity (S, columns) values for untreated and treated Z30. Conversion (triangle), S(*p*-xylene) (black), S(2,5-hexanedione) (HDO, grey), S(3-methyl-2-cyclopentenone) (MCP, white), S(alkylated products) (AP, square pattern), and S(higher aromatics) (HA, diagonal pattern). Reaction conditions: 180 °C, 40 bar total pressure (pressurised at 180 °C), 5.5 mL DMF, 0.45 mL tridecane (internal standard), 16.5 mL hexane, 110 mg catalyst, 20 h. The carbon mass balance was >96.5%, except in case of Z30.2 it was 85%.

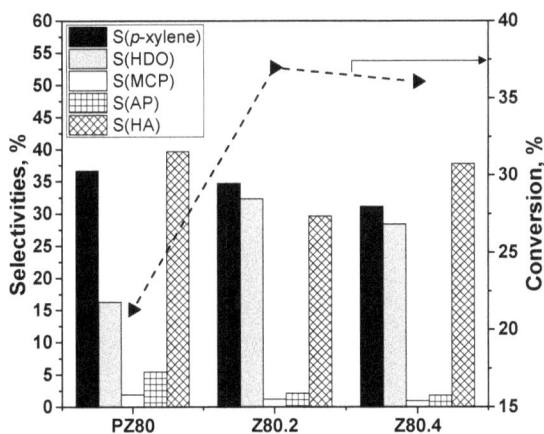

Figure 9. Conversion (X, symbol + line) and selectivity (S, columns) values for untreated and treated Z80. Conversion (triangle), S(*p*-xylene) (black), S(2,5-hexanedione) (HDO, grey), S(3-methyl-2-cyclopentenone) (MCP, white), S(alkylated products) (AP, square pattern) and S(higher aromatics) (HA, diagonal pattern). Reaction conditions: 180 °C, 40 bar total pressure (pressurised at 180 °C), 5.5 mL DMF, 0.45 mL tridecane (internal standard), 16.5 mL hexane, 110 mg catalyst, 20 h. The carbon mass balance was in all cases >94%.

This would suggest that increased porosity promotes the transformation of DMF to reaction products. However, the additional desilication and increase in the porosity for both Z30 and Z80 does not promote further the conversion when compared to the 0.2 M NaOH-treated samples. Nonetheless, the activity of the Z30.4 and Z80.4 increased to 27.1% and 36.1% when compared to the PZ30 and PZ80, respectively. As the effect of the alkaline treatment was more pronounced in the case of Z80, this material was selected for the catalytic tests at high temperature. It can be seen that when the reaction temperature is increased to 250 °C, the DMF conversion for Z80.2 is significantly enhanced from 16% to 51% when compared to PZ80 (Figure 10) under the same conditions.

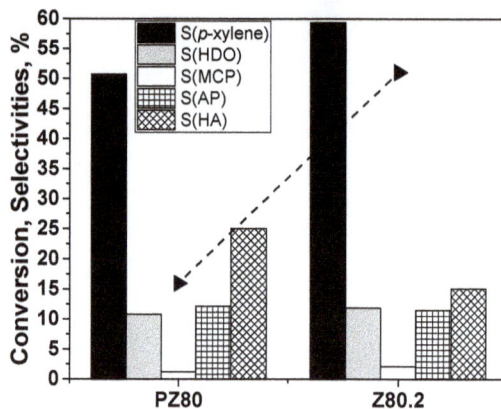

Figure 10. Conversion (X, symbol + line) and selectivity (S, columns) values for untreated PZ80 and treated Z80.4. Conversion (triangle), S(*p*-xylene) (black), S(2,5-hexanedione) (HDO, grey), S(3-methyl-2-cyclopentenone) (MCP, white), S(alkylated products) (AP, square pattern) and S(higher aromatics) (HA, diagonal pattern). Reaction conditions: 250 °C, 55 bar total pressure (pressurised at 250 °C), 5.5 mL DMF, 0.45 mL tridecane (internal standard), 16.5 mL hexane, 110 mg catalyst, 20 h. The carbon mass balance was in all cases >95%.

2.6.2. Products Selectivity

Chang and co-workers described the evolution of products selectivities as a function of time over the H-BEA catalyst and found that the selectivity to *p*-xylene increases in an almost linear fashion with the increasing time/conversion and so does the selectivity to 2,5-hexanedione. The selectivity to hexanedione, however, starts to decrease as a result of hexanedione being converted back to DMF when certain DMF conversion is reached (60%) [9]. As it can be seen from Figure 7, hexanedione selectivity starts to decline in our case at 40–50% DMF conversion to get dehydrated back to DMF to be converted to *p*-xylene. It can be seen from Figures 8 and 9 that at 180 °C the S(*p*-xylene) increased with increasing conversion only for PZ30 (43.2%) to Z30.2 (44.1%), while it decreased in all other cases of both Z30 and Z80.

It was suggested that this decrease in selectivity to *p*-xylene could be caused by the retention of formed *p*-xylene on acid sites, which would react further to alkylated products (AP) and higher aromatics (HA), which further react to form unrecoverable polymerised products (coke) [32]. However, quantum mechanics calculations showed both hexanedione and oxanorbornene intermediate adsorb stronger (ca. 25.8 kcal/mol) to the acid site as compared with DMF (17.9 kcal/mol) and *p*-xylene (16.6 kcal/mol) [12]. Oxanorbornene can be dehydrated by Brønsted acid sites, as it has been demonstrated by several authors [7,33]; however, Lewis acid sites can be active for the dehydration step as well [13,15,34]. As the adsorbed 7-oxabicyclo[2.2.1]hept-2-ene intermediate can be easily transformed into *p*-xylene, DMF and hexanedione were indicated as the most probable causes for the formation of oligomeric byproducts.

2.6.3. Increasing the Reaction Temperature and Catalyst Reusability

Z80 was chosen as a representative for testing at higher temperatures (Figure 10). Even after the significant conversion increases for Z80.2 (51%) vs PZ80 (16%), S(*p*-xylene) is still higher than at lower conversion (PZ80) in this case, which further corroborates the hypothesis that the higher reaction temperature is beneficial for the prevention of the unwanted oxanorbornene intermediate reactivity to alkylation and condensation byproducts. Furthermore, the behaviour of Z80.2 at the same temperature of 250 °C but at lower pressure ethylene was tested (Table S1, entries 8, 9). It shows that both the conversion of DMF and S(p-xylene) drop almost to the reactivity of the parent PZ80 when lower ethylene pressure is used. This supports the hypothesis that high pressure of ethylene is needed to increase the probability of the cycloaddition step as it was also suggested by Nikbin and co-workers [7].

We also carried out a catalyst reusability test with Z80.2 to investigate how the catalyst performs in repeated tests (Table 3).

Table 3. Catalyst Z80.2 reusability results in the Diels–Alder cycloaddition of 2,5-dimethylfuran and ethylene. Reaction conditions: 220 °C, 55 bar total pressure, 1100 rpm, 2 h reaction time, DMF (12 mL), tridecane (0.98 mL), dodecane (25 mL), 180 mg of Z80.2.

Test Description	X	S(p-Xylene)	S(HDO)	S(MCP)	S(AP)	S(HA)	Carbon Mass Balance
fresh	42.9	50.1	13.9	0.8	16.0	19.2	93.2
reuse 1 *	41.4	51.6	13.7	0.5	17.9	16.3	95.5
reuse 2	42.4	48.8	14.5	0.6	16.0	20.2	92.6
reuse 3	41.7	48.9	14.7	0.5	14.8	21.2	94.6

* catalyst only washed and dried catalyst before the test, catalyst for further reuses (2, 3) was washed, dried, and calcined at 550 °C for 6 h. X, conversion; HDO, hexanedione; MCP, 3-methyl-2-cyclopentenone; AP, alkylated products; and HA, higher aromatics.

It can be seen that the catalyst is reusable without the loss of activity in three subsequent reuses. Also, it should be noted that before the first reuse, the catalyst (pale brown colour) was only washed and dried and still performed similarly to the fresh catalyst (white colour) and last two reuses (white catalyst colour).

2.6.4. Mesoporosity vs Acidity

Both Brønsted [14,35] and Lewis [13,16,34,36] acid sites were shown to be suitable for tandem Diels–Alder cycloaddition of furanic compounds and ethylene [3,18,37]. According to calculations, Brønsted acidity was proven to be more effective than Lewis in the dehydration of the oxanorbornene intermediate [18]. We found that the conversion of DMF per Lewis acid site linearly increased with the increase in the mesopore surface area (Figure 11).

This could mean that as more mesopores are created, the availability of Lewis acid sites increases. We could not find any correlation with Brønsted acid sites, which can be explained either by the Diels–Alder cycloaddition step being rate-determining [18] or also because of the possible synergy between the Brønsted and Lewis sites as it was documented previously [38,39]. In our catalysts, we observed Lewis and Brønsted acid sites and—as the trend of Brønsted acidity was not apparent—the effect of Brønsted acidity was inconclusive. It was suggested in the literature a long time ago that Lewis acids significantly enhance the rate of the Diels–Alder cycloaddition [5,40,41]. However, this depends on the reaction conditions, such as temperature and pressure, as well as concentration of acid sites (high/low), which was shown to affect the regime kinetics [12,13]. As in our case we have dual acidity (Brønsted and Lewis), the most plausible explanation is the enhancement of reactivity by desilication by creating extra-framework Lewis sites (as documented by NMR) with better accessibility, whereas the role of the Brønsted acid sites deserves further investigation.

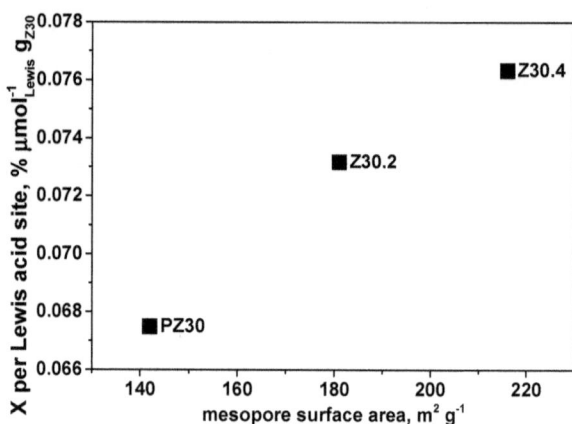

Figure 11. Conversion per Lewis acid site for untreated and treated Z30 vs mesopore surface area for cycloaddition reaction of DMF with ethylene. Reaction conditions: 180 °C, 40 bar total pressure (pressurised at 180 °C), 5.5 mL DMF, 0.45 mL tridecane (internal standard), 16.5 mL hexane, 110 mg catalyst, 20 h.

PZ30 and PZ80 showed similar conversion of DMF of 19.3% and 21.3% and also similar mesopore surface area of 142 and 131 m^2 g^{-1}, respectively. Still, PZ30 displayed 51% higher selectivity towards *p*-xylene. We believe that this is due to the fact that PZ30 possesses more strong acid sites than PZ80 (416 vs 278 μmol g^{-1}, respectively). This acidity is required for the dehydration of the cycloadduct intermediate to *p*-xylene (Scheme 1). It is also true that there are more Brønsted/Lewis acid sites in PZ30 (295/286 μmol g^{-1}) vs PZ80 (166/167 μmol g^{-1}), which is expected based on the Si/Al ratio. NH$_3$-TPD measurements confirm that PZ30 and Z30.4 have almost identical amounts of strong acid sites of 416 and 426 μmol g^{-1}, respectively. However, Z30.4 exhibits a 40% increase in conversion compared to its parent PZ30. We propose that this is a result of the increased mesopore area in Z30.4, which was found to be 52% higher when compared to PZ30 (Table 1). This is also supported in the case of alkaline-treated Z80. PZ80 and Z80.4 showed similar amounts of strong acid sites of 278 and 244 μmol g^{-1}, respectively. At the same time, Z80.4 displayed 66% conversion enhancement in comparison with PZ80. Similarly, Z80.4 possessed double the mesopore surface area of PZ80 (Table 1). The increase in conversion for both treated Z30 and Z80 is likely due to the increase in mesoporosity, which could involve both a) higher tolerance to coke formation than in microporous samples thereby extending catalyst lifetime and b) improved reactants diffusion. Coke tolerance of Z80.2 is supported by the reusability test, where the catalyst with coke deposited (pale brown colour) performed the same as the fresh catalyst (white colour) (Table 3).

3. Materials and Methods

3.1. Materials

Commercial ZSM-5 (MFI) zeolites were obtained in the ammonium form (CBV 3024E, SiO$_2$/Al$_2$O$_3$ = 30 and CBV 8014, SiO$_2$/Al$_2$O$_3$ = 80) from Zeolyst International Inc. (Conshohocken, PA, USA). 2,5-dimethylfuran (99%), ammonium nitrate, and tridecane were obtained from Sigma Aldrich (St. Louis, MO, USA). Sodium hydroxide pellets and hexane were obtained from Fisher Scientific (Hampton, NH, USA), and apart from DMF, were used without further purification. DMF was purified by vacuum distillation.

3.2. Catalyst Preparation

Conventional ZSM-5. The ammonium form of ZSM-5 was treated at 550 °C in static air at a heating ramp of 1 °C min^{-1} for 6 h to produce the proton form of the same material. This was repeated for subsequent calcinations. The parent samples are noted as P (e.g., PZ30).

Alkaline treated ZSM-5. The alkaline-treated ZSM-5 samples are noted according to the amount of NaOH used (e.g., Z30.2 and Z30.4 for 0.2 M NaOH and 0.4 M NaOH treatment of Z30, respectively). The desilication procedure was performed by treating the ammonium form of the zeolite with aqueous sodium hydroxide (30 mL g^{-1}). The mixture was stirred for 30 min at 65 °C and then quenched in an ice bath to prevent further desilication. The slurry was filtered using Whatman grade 3 filter paper. The sample was washed and then dried overnight. The desilicated catalyst was converted to the ammonium form by treating the sample with 1 M ammonium nitrate for 120 min at 80 °C (20 mL g^{-1}). The suspension was cooled, then filtered and dried overnight. Finally, the zeolite was converted from the ammonium into the proton form by calcination as described above.

3.3. Catalytic Activity

DMF was distilled beforehand to remove heavy contaminants, leaving behind a brown residue. The reaction of 2,5-DMF and ethylene was carried out in a 45 mL closed Parr model 4717 pressure reactor. DMF (5.5 mL), tridecane (0.45 mL), hexane (16.5 mL), and 110 mg of catalyst were charged into the vessel. The vessel was stirred using a cross-shaped magnetic stir bar at 1100 rpm, purged with nitrogen, and heated to 180 °C using an aluminium block and a hotplate. As soon as the vessel reached the required temperature, the vessel was pressurised to a total pressure of 40 bar with ethylene for 20 h. At the end of the reaction, the vessel was quenched in an ice bath. The liquid phase was separated from the spent catalyst by centrifugation. Product identification was carried out using gas chromatography (Agilent series 7820A, HP-5 capillary column, 30 m, 0.32 mm, 0.25 μm) (Agilent Technologies, Santa Clara, CA, USA) using flame ionisation detector. The conversion and selectivities were calculated based on the below equations:

$$\text{Conversion}_{\text{DMF}}\ (\%) = \frac{[\text{DMF}]_{\text{initial}} - [\text{DMF}]_{\text{end}}}{[\text{DMF}]\text{initial}} \times 100$$

$$\text{Selectivity}_{\text{product}}\ (\%) = \frac{[\text{product}]}{\text{sum of } [\text{products}]} \times 100$$

Carbon mass balance was a sum of concentrations of all identified products and unreacted DMF. Reusability tests were performed with Z80.2 at the following reaction conditions. DMF (12 mL), tridecane (0.98 mL), dodecane (25 mL), and 180 mg of Z80.2 were added to a 75 mL stainless-steel vessel and purged with nitrogen. The vessel was heated to 220 °C and pressurised to a total pressure of 55 bar with ethylene. The reaction was carried out and stirred at 1100 rpm for 2 h. After the reaction was allowed to cool, the spent catalyst was filtered and washed with ethanol and dried in an oven overnight at 110 °C. The spent catalyst was then calcined as described in Section 3.2. The first reuse test was carried out with only washed and dried catalyst (turned pale brown), whereas the subsequent two reuses were done with catalyst regenerated by calcination.

3.4. Characterisation

X-ray diffraction was carried out using a Panalytical X'Pert PRO HTS X-ray diffractometer (PANAlytical, Almelo, The Netherlands) using Cu Kα radiation (λ = 0.154 nm). Data were recorded in the 2θ range of 5–50°. Scanning electron microscopy (SEM) was carried out using Hitachi S-4800 Field Emission Scanning Electron Microscope (Hitachi, Tokyo, Japan). For ICP-OES measurements, approximately 1 mg of zeolite was sonicated in 1 mL of water until a suspension was obtained. The amount of Si and Al in the suspension was measured and used to calculate the SiO$_2$/Al$_2$O$_3$ ratios. Each value of Si and Al was measured three times and averaged. Nitrogen sorption isotherms were measured using Micromeritics ASAP 2020 (Micromeritics Instrument Corporation, Norcross, Georgia, USA) at 77 K. The surface area was calculated using the BET equation with the P/P_0 range 0.05–0.30. Total pore volume was taken as a volume at $P/P_0 = 0.99$. The micropore volume was calculated using the NLDFT method up to 2 nm pore size. Mesopore surface area was calculated as the difference between the BET surface area and the micropore volume. Temperature desorption of ammonia (NH_3-TPD) was carried out using Micromeritics Autochem II 2920 (Micromeritics Instrument Corporation, Norcross, GA, USA) equipped with a thermal conductivity detector. The sample (50 mg) was pre-treated in air at 550 °C at a rate of 5 °C min^{-1} and held for 60 min. Then the gas flow was changed from air to helium and flowed at a rate of 20 cm^3 min^{-1} for 20 min. The sample was then cooled to 150 °C. Then, 5% NH_3/He mixture was flowed at a rate of 20 cm^3 min^{-1} for 30 min. The gas was changed to helium and flowed at a rate of 20 cm^3 min^{-1} for 30 min. For TPD, the sample was heated to 700 °C at a rate of 10 °C min^{-1}. Diffuse reflectance infrared Fourier transform (DRIFT) spectra of catalysts were taken on a Nicolet NEXUS FTIR spectrometer (Thermo Fisher Scientific, Waltham, MA, USA) using powdered catalyst mixtures with KBr. H-ZSM-5 was mixed with KBr in a 4:1 weight ratio. The powder was pre-treated at 150 °C/0.01 kPa for 60 min under vacuum. The sample was dried under a flow of nitrogen. Pyridine was dropped onto the sample. The samples were exposed to pyridine for 60 min and then degassed at 150 °C to remove physisorbed pyridine. The Brønsted (B) and Lewis (L) sites were determined using peak intensities at 1540 cm^{-1} and 1450 cm^{-1}, respectively. ^{27}Al MAS NMR spectra were recorded using a Varian VNMRS spectrometer at 104.2 MHz at room temperature in 4 mm rotor with a spinning rate of 14,000 Hz with "onepul" pulse sequence, direct excitation, 1 µs pulse duration, 10 ms acquisition time, 200 ms recycle time, collected 7000 transients, spectral width 416.7 kHz.

4. Conclusions

We have demonstrated that alkaline treatment of H-ZSM-5 is beneficial for the [4 + 2] Diels–Alder cycloaddition of DMF and ethylene when compared to the untreated parent. The alkaline treatment results in the preferential silicon dissolution, which leads to enhanced mesoporosity and also increased conversion of DMF. Furthermore, we showed that by tuning the alkaline treatment one can produce mesoporosity while preserving the crystallinity of the zeolite. This modification of H-ZSM-5 could be applied to other zeolites for subsequent dehydration reactions involving larger aromatic molecules.

The correlation between the increase of the surface area of the mesopores and activity per Lewis acid site was found, which was caused by the better availability of the extra-framework Lewis acid sites in the desilicated zeolite. However, although no correlation was found for Brønsted acidity, its participation cannot be neglected and deserves further investigation.

Also, elevated temperatures are beneficial to prevent further polymerisation reactions, and increased ethylene pressure can significantly enhance the rate of the cycloaddition reaction step.

Supplementary Materials: The following are available online at http://www.mdpi.com/2073-4344/8/6/253/s1, Figure S1: NH_3-TPD profile (left) and column view of number of acid sites (right) of untreated and treated Z30. Adsorption of NH_3 was performed for 30 min at 150 °C followed by inert flush and TPD up to 700 °C. Columns legend: total acidity (black), weak acid sites (pattern), and strong acid sites (grey), Figure S2: NH_3-TPD profile (left) and column view of number of acid sites (right) of untreated and treated Z80. Adsorption of NH_3 was performed for 30 min at 150 °C followed by inert flush and TPD up to 700 °C. Columns legend: total acidity (black), weak acid sites (pattern), and strong acid sites (grey), Figure S3: SEM micrographs of PZ30 (A–C) and Z30 (D–F), Figure S4 SEM micrographs of PZ80 (A–C) and Z80.2 (D–F), Table S1: Results of Diels–Alder catalytic testing of DMF and ethylene over untreated and treated H-ZSM-5. Reaction conditions: 5.5 mL DMF, 0.45 mL tridecane (internal standard), 16.5 mL hexane, 110 mg catalyst, 20 h. Conversions and selectivities in %. HDO is 2,5-hexanedione, MCP is 3-methyl-2-cyclopentenone, AP are alkylated products, HA are higher aromatics. The rest is selectivity to unknown products. The carbon mass balance was in all cases >94%.

Author Contributions: Conceptualization, J.A.L.-S. and P.P.; Methodology, J.M. and P.P.; Validation, J.M., L.M., P.P., L.D.V., and J.A.L.-S.; Formal Analysis, J.M.; Investigation, J.M., L.D.V., and L.M.; Resources, J.A.L.-S.; Data Curation, J.M. and P.P.; Original Draft Preparation, J.M.; Review and Editing Manuscript, P.P., L.M., L.D.V., and J.A.L.-S.; Visualization, J.M.; Supervision, J.A.L.-S. and P.P.; Project Administration, J.A.L.-S. and P.P.; Funding Acquisition, J.A.L.-S.

Funding: This research was funded by the Engineering and Physical Sciences Research Council (EPSRC) grant number EP/K014773/1. Liqaa Majdal wishes to acknowledge funding of this research by the Higher Committee for Education Development in Iraq.

Acknowledgments: The authors thank Ivan Kozhevnikov's group especially Hossein Bayahia for their help with the FTIR-pyridine measurements and Rob Clowes for his help with the BET measurements. The authors would also like to thank David Apperley for the measurement of solid-state NMR spectra, which were obtained at the EPSRC UK National Solid-State NMR Service at Durham. The authors acknowledge the staff and use of the MicroBioRefinery facility (financed by the Department of Business Skills and Innovation (Regional Growth Fund), where all the experiments were performed.

Conflicts of Interest: The authors declare no conflict of interest. The funding sponsors had no role in the design of the study; in the collection, analyses, or interpretation of data; in the writing of the manuscript; or in the decision to publish the results.

References

1. Lin, Z.; Ierapetritou, M.; Nikolakis, V. Aromatics from lignocellulosic biomass: Economic analysis of the production of p-xylene from 5-hydroxymethylfurfural. *AIChE J.* **2013**, *59*, 2079–2087. [CrossRef]
2. Maneffa, A.; Priecel, P.; Lopez-Sanchez, J.A. Biomass-derived renewable aromatics: Selective routes and outlook for p-xylene commercialisation. *ChemSusChem* **2016**, *9*, 2736–2748. [CrossRef] [PubMed]
3. Settle, A.E.; Berstis, L.; Rorrer, N.A.; Roman-Leshkov, Y.; Beckham, G.T.; Richards, R.M.; Vardon, D.R. Heterogeneous diels-alder catalysis for biomass-derived aromatic compounds. *Green Chem.* **2017**, *19*, 3468–3492. [CrossRef]
4. Cheng, Y.T.; Huber, G.W. Production of targeted aromatics by using diels-alder classes of reactions with furans and olefins over zsm-5. *Green Chem.* **2012**, *14*, 3114–3125. [CrossRef]
5. Brion, F. On the lewis acid catalyzed diels-alder reaction of furan. Regio- and stereospecific synthesis of substituted cyclohexenols and cyclohexadienols. *Tetrahedron Lett.* **1982**, *23*, 5299–5302. [CrossRef]
6. Williams, C.L.; Chang, C.C.; Do, P.; Nikbin, N.; Caratzoulas, S.; Vlachos, D.G.; Lobo, R.F.; Fan, W.; Dauenhauer, P.J. Cycloaddition of biomass-derived furans for catalytic production of renewable p-xylene. *ACS Catal.* **2012**, *2*, 935–939. [CrossRef]
7. Nikbin, N.; Do, P.T.; Caratzoulas, S.; Lobo, R.F.; Dauenhauer, P.J.; Vlachos, D.G. A dft study of the acid-catalyzed conversion of 2,5-dimethylfuran and ethylene to p-xylene. *J. Catal.* **2013**, *297*, 35–43. [CrossRef]
8. Shiramizu, M.; Toste, F.D. On the diels-alder approach to solely biomass-derived polyethylene terephthalate (pet): Conversion of 2,5-dimethylfuran and acrolein into p-xylene. *Chem.-Eur. J.* **2011**, *17*, 12452–12457. [CrossRef] [PubMed]
9. Chang, C.C.; Green, S.K.; Williams, C.L.; Dauenhauer, P.J.; Fan, W. Ultra-selective cycloaddition of dimethylfuran for renewable p-xylene with h-bea. *Green Chem.* **2014**, *16*, 585–588. [CrossRef]
10. Brandvold, T. Carbohydrate Route to Para-Xylene and Terephthalic Acid. U.S. Patent 8,314,267, 20 November 2012.
11. Roman-Leshkov, Y.; Barrett, C.J.; Liu, Z.Y.; Dumesic, J.A. Production of dimethylfuran for liquid fuels from biomass-derived carbohydrates. *Nature* **2007**, *447*, U982–U985. [CrossRef] [PubMed]
12. Patet, R.E.; Nikbin, N.; Williams, C.L.; Green, S.K.; Chang, C.-C.; Fan, W.; Caratzoulas, S.; Dauenhauer, P.J.; Vlachos, D.G. Kinetic regime change in the tandem dehydrative aromatization of furan diels–alder products. *ACS Catal.* **2015**, *5*, 2367–2375. [CrossRef]
13. Rohling, R.Y.; Uslamin, E.; Zijlstra, B.; Tranca, I.C.; Filot, I.A.W.; Hensen, E.J.M.; Pidko, E.A. An active alkali-exchanged faujasite catalyst for p-xylene production via the one-pot diels–alder cycloaddition/dehydration reaction of 2,5-dimethylfuran with ethylene. *ACS Catal.* **2018**, *8*, 760–769. [CrossRef] [PubMed]
14. Kim, J.-C.; Kim, T.-W.; Kim, Y.; Ryoo, R.; Jeong, S.-Y.; Kim, C.-U. Mesoporous mfi zeolites as high performance catalysts for diels-alder cycloaddition of bio-derived dimethylfuran and ethylene to renewable p-xylene. *Appl. Catal. B Environ.* **2017**, *206*, 490–500. [CrossRef]
15. Patet, R.E.; Caratzoulas, S.; Vlachos, D.G. Tandem aromatization of oxygenated furans by framework zinc in zeolites. A computational study. *J. Phys. Chem. C* **2017**, *121*, 22178–22186. [CrossRef]

16. Patet, R.E.; Fan, W.; Vlachos, D.G.; Caratzoulas, S. Tandem diels–alder reaction of dimethylfuran and ethylene and dehydration to para-xylene catalyzed by zeotypic lewis acids. *ChemCatChem* **2017**, *9*, 2523–2535. [CrossRef]

17. Wijaya, Y.P.; Suh, D.J.; Jae, J. Production of renewable p-xylene from 2,5-dimethylfuran via diels-alder cycloaddition and dehydrative aromatization reactions over silica-alumina aerogel catalysts. *Catal. Commun.* **2015**, *70*, 12–16. [CrossRef]

18. Nikbin, N.; Feng, S.; Caratzoulas, S.; Vlachos, D.G. P-xylene formation by dehydrative aromatization of a diels–alder product in lewis and brønsted acidic zeolites. *J. Phys. Chem. C* **2014**, *118*, 24415–24424. [CrossRef]

19. Williams, C.L. Production of Sustainable Aromatics from Biorenewable Furans. Doctoral Dissertations, University of Massachusetts, Amherst, MA, USA, 2014.

20. Moliner, M.; Martinez, C.; Corma, A. Multipore zeolites: Synthesis and catalytic applications. *Angew. Chem. Int. Ed.* **2015**, *54*, 3560–3579. [CrossRef] [PubMed]

21. Kim, J.; Choi, M.; Ryoo, R. Effect of mesoporosity against the deactivation of mfi zeolite catalyst during the methanol-to-hydrocarbon conversion process. *J. Catal.* **2010**, *269*, 219–228. [CrossRef]

22. Groen, J.C.; Peffer, L.A.A.; Moulijn, J.A.; Perez-Ramirez, J. Mesoporosity development in zsm-5 zeolite upon optimized desilication conditions in alkaline medium. *Colloid Surf. A* **2004**, *241*, 53–58. [CrossRef]

23. Dean, A.Y. Hydrocarbon Conversion Process and Catalyst Comprising a Crystalline Alumino-Silicate Leached with Sodium Hydroxide. US3326797 A, 20 June 1967.

24. Verboekend, D.; Perez-Ramirez, J. Design of hierarchical zeolite catalysts by desilication. *Catal. Sci. Technol.* **2011**, *1*, 879–890. [CrossRef]

25. Groen, J.C.; Moulijn, J.A.; Perez-Ramirez, J. Decoupling mesoporosity formation and acidity modification in zsm-5 zeolites by sequential desilication-dealumination. *Microporous Mesoporous Mater.* **2005**, *87*, 153–161. [CrossRef]

26. Brunner, E.; Pfeifer, H.; Auroux, A.; Lercher, J.; Jentys, A.; Brait, A.; Garrone, E.; Fajula, F. *Acidity and Basicity*; Springer: Berlin/Heidelberg, Germany, 2008.

27. Rodriguez-Gonzalez, L.; Hermes, F.; Bertmer, M.; Rodriguez-Castellon, E.; Jimenez-Lopez, A.; Simon, U. The acid properties of h-zsm-5 as studied by nh3-tpd and al-27-mas-nmr spectroscopy. *Appl. Catal. A Gen.* **2007**, *328*, 174–182. [CrossRef]

28. Rac, V.; Rakic, V.; Miladinovic, Z.; Stosic, D.; Auroux, A. Influence of the desilication process on the acidity of hzsm-5 zeolite. *Thermochim. Acta* **2013**, *567*, 73–78. [CrossRef]

29. Sadowska, K.; Wach, A.; Olejniczak, Z.; Kustrowski, P.; Datka, J. Hierarchic zeolites: Zeolite zsm-5 desilicated with naoh and naoh/tetrabutylamine hydroxide. *Microporous Mesoporous Mater.* **2013**, *167*, 82–88. [CrossRef]

30. Jiang, Y.J.; Huang, J.; Dai, W.L.; Hunger, M. Solid-state nuclear magnetic resonance investigations of the nature, property, and activity of acid sites on solid catalysts. *Solid State Nucl. Mag.* **2011**, *39*, 116–141. [CrossRef] [PubMed]

31. Sklenak, S.; Dedecek, J.; Li, C.; Wichterlova, B.; Gabova, V.; Sierka, M.; Sauer, J. Aluminium siting in the zsm-5 framework by combination of high resolution al-27 nmr and dft/mm calculations. *Phys. Chem. Chem. Phys.* **2009**, *11*, 1237–1247. [CrossRef] [PubMed]

32. Do, P.T.M.; McAtee, J.R.; Watson, D.A.; Lobo, R.F. Elucidation of diels-alder reaction network of 2,5-dimethylfuran and ethylene on hy zeolite catalyst. *ACS Catal.* **2013**, *3*, 41–46. [CrossRef] [PubMed]

33. Salavati-fard, t.; Caratzoulas, S.; Doren, D.J. Solvent effects in acid-catalyzed dehydration of the diels-alder cycloadduct between 2,5-dimethylfuran and maleic anhydride. *Chem. Phys.* **2017**, *485–486*, 118–124. [CrossRef]

34. Chang, C.-C.; Je Cho, H.; Yu, J.; Gorte, R.J.; Gulbinski, J.; Dauenhauer, P.; Fan, W. Lewis acid zeolites for tandem diels-alder cycloaddition and dehydration of biomass-derived dimethylfuran and ethylene to renewable p-xylene. *Green Chem.* **2016**, *18*, 1368–1376. [CrossRef]

35. Li, Y.-P.; Head-Gordon, M.; Bell, A.T. Computational study of p-xylene synthesis from ethylene and 2,5-dimethylfuran catalyzed by h-bea. *J. Phys. Chem. C* **2014**, *118*, 22090–22095. [CrossRef]

36. Wijaya, Y.P.; Kristianto, I.; Lee, H.; Jae, J. Production of renewable toluene from biomass-derived furans via diels-alder and dehydration reactions: A comparative study of lewis acid catalysts. *Fuel* **2016**, *182*, 588–596. [CrossRef]

Catalysts **2018**, *8*, 253

37. Fan, W.; Cho, H.J.; Ren, L.; Vattipailli, V.; Yeh, Y.-H.; Gould, N.G.; Xu, B.; Gorte, R.J.; Lobo, R.; Dauenhauer, P.J.; et al. Renewable p-xylene from 2,5-dimethylfuran and ethylene using phosphorus-containing zeolite catalysts. *ChemCatChem* **2016**. [CrossRef]

38. Yu, Z.; Li, S.; Wang, Q.; Zheng, A.; Jun, X.; Chen, L.; Deng, F. Brønsted/lewis acid synergy in h–zsm-5 and h–mor zeolites studied by 1h and 27al dq-mas solid-state nmr spectroscopy. *J. Phys. Chem. C* **2011**, *115*, 22320–22327. [CrossRef]

39. Zheng, A.; Li, S.; Liu, S.-B.; Deng, F. Acidic properties and structure–activity correlations of solid acid catalysts revealed by solid-state nmr spectroscopy. *Acc. Chem. Res.* **2016**, *49*, 655–663. [CrossRef] [PubMed]

40. Houk, K.N.; Strozier, R.W. Lewis acid catalysis of diels-alder reactions. *J. Am. Chem. Soc.* **1973**, *95*, 4094–4096. [CrossRef]

41. Birney, D.M.; Houk, K.N. Transition structures of the lewis acid-catalyzed diels-alder reaction of butadiene with acrolein—The origins of selectivity. *J. Am. Chem. Soc.* **1990**, *112*, 4127–4133. [CrossRef]

catalysts

MDPI

Communication

Bio-Glycidol Conversion to Solketal over Acid Heterogeneous Catalysts: Synthesis and Theoretical Approach

Maria Ricciardi [1], Laura Falivene [2,*], Tommaso Tabanelli [3], Antonio Proto [1], Raffaele Cucciniello [1,*] and Fabrizio Cavani [3]

[1]   Department of Chemistry and Biology, "Adolfo Zambelli" University of Salerno, Via Giovanni Paolo II, 132, 84084 Fisciano (SA), Italy; mricciardi@unisa.it (M.R.); aproto@unisa.it (A.P.)
[2]   KAUST Catalysis Center (KCC), King Abdullah University of Science and Technology (KAUST), Thuwal 23955-6900, Saudi Arabia
[3]   Department of Industrial Chemistry "Toso Montanari", Alma Mater Studiorum Università di Bologna, Viale del Risorgimento 4, 40136 Bologna, Italy; tommaso.tabanelli@unibo.it (T.T.); fabrizio.cavani@unibo.it (F.C.)
*   Correspondence: laura.falivene@kaust.edu.sa (L.F.); rcucciniello@unisa.it (R.C.); Tel.: +39-33-3348-3532 (L.F.); +39-089-969-366 (R.C.)

Received: 29 August 2018; Accepted: 7 September 2018; Published: 11 September 2018

Abstract: The present work deals with the novel use of heterogeneous catalysts for the preparation of solketal from bio-glycidol. Sustainable feedstocks and mild reaction conditions are considered to enhance the greenness of the proposed process. Nafion NR50 promotes the quantitative and selective acetalization of glycidol with acetone. DFT calculations demonstrate that the favored mechanism consists in the nucleophilic attack of acetone to glycidol concerted with the ring opening assisted by the acidic groups on the catalyst and in the following closure of the five member ring of the solketal.

Keywords: biomass; glycidol; heterogeneous catalysis; solketal

1. Introduction

Nowadays with the depletion of fossil fuels many efforts are devoted to the development of new green routes to convert renewables into biofuels [1,2]. This objective fully addresses the Green Chemistry principles proposed by Anastas and Warner in 1998 [3]. Among the others, the conversion of glycerol, mainly obtained as by product during biodiesel production, into value-added products is extremely important [4]. To this extent, several strategies have been investigated to convert glycerol into propanediols, dihydroxyacetone, allyl alcohol, polyglycerols, glycerol ethers, glycerol esters, etc. [5–9]. Among all the considered routes, the preparation of cyclic acetals and ketals through the reaction between glycerol and aldehydes/ketones in the presence of an acid catalyst represents one of the most promising alternatives [10,11]. In details, the condensation of glycerol with acetone yields a very interesting compound, namely solketal (2,2-dimethyl-1,3-dioxolane-4-methanol), employed as flavoring agent, surfactant and fuel additive. Herein water is produced as by-product and need to be removed to hinder the reversibility of the reaction. Solketal can be directly used to reduce biodiesel viscosity and to satisfy the established values for flash point and oxidation stability [12]. The most diffused approach for the synthesis of solketal starting from glycerol requires the use of large amounts of a strong homogeneous Bronsted acid catalyst. Recently, several papers reported on the use of heterogenous catalysts like Amberlyst resins, zeolites, montmorillonite K10, sulfonated silicas and silica-supported heteropolyacids. In 2012, Pescarmona and coworkers described the promising application of heterogeneous Lewis acid catalysts for the conversion of glycerol to solketal [13].

As alternative to glycerol, glycidol (2,3-epoxy-1-propanol) can be considered a potential candidate as starting molecule to synthesize solketal. However, the preparation of solketal starting from glycidol was barely investigated to date and always in the presence of homogeneous catalytic systems. More in detail, Iranpoor and Kazemi reported the conversion of glycidol to solketal (89% isolated yield after 2 h of reaction) in the presence of 0.2 molar equivalent of $RuCl_3$ in refluxing acetone [14]. Afterwards, the same research group reported good results also using 0.2 molar equivalent of iron(III)trifluoroacetate in refluxing acetone (89% as isolated yield after 4 h of reaction) [15]. More recently, Procopio et al. showed the quantitative conversion of glycidol to solketal in acetone at room temperature after 48 h in the presence of 1% in moles of $Er(OTf)_3$ [16]. The authors suggested a mechanistic scenario involving the oxirane ring activation through the coordination to the Er(III) followed by the nucleophilic attack of acetone.

The use of glycidol as starting material for solketal preparation becomes more interesting in the light of the recently investigated bio-based routes for its preparation. More in detail, we recently described the preparation of glycidol through the conversion of 2-chloro-1,3-propanediol (β-MCH), a by-product in the bio-based epichlorohydrin production plant [4,17,18]. This approach allows to valorize the entire production chain to bio-epichlorohydrin minimizing the production of waste in agreement with the twelve principles of Green Chemistry.

As a matter of fact, glycidol in turn can be easily used as starting material to produce high-value products through catalysis [19].

In this work, we report for the first time the selective preparation of solketal through glycidol ketalization with acetone in the presence of acid heterogeneous catalysts with the aim to increase the sustainability of this process.

The effect of temperature, glycidol/acetone ratio and catalyst loading has been investigated to find the optimal conditions for the reaction. Moreover, other ketones have been tested to prove the versatility of this approach and extend its generality. Finally, DFT calculations have been performed in order to rationalize the mechanism occurring.

2. Results and Discussion

2.1. Glycidol Conversion to Solketal: Reaction Conditions Optimization

Initially, the tests were performed using a catalyst loading of 10 wt % with respect to glycidol and a glycidol/acetone molar ratio of 43, heating the system to reflux, as reported in literature for glycidol acetalization in homogeneous phase [20]. In these conditions, acetone acts both as reagent and reaction solvent, avoiding the need of any other organic solvent, finally simplifying the purification of the products and acetone recovery and recycle. This represent an important aspect for a potential industrial scale-up [21]. Herein, the preparation of solketal starting from glycidol permits us to easily separate the desired product at the end of the process using a rotary evaporator under reduced pressure thanks to the highly different boiling points of solketal (188 °C) and acetone (56 °C).

Several heterogeneous catalysts (both Lewis and Brønsted acids) have been used to promote glycidol acetalization to solketal. Nafion NR50, Montomorillonite K10 and Amberlyst-15 are commercially available whereas sulfonated charcoal, sufonated mesoporous silica and supported metal triflates have been prepared and successfully employed in acid-demanding processes [18]. As shown in Table 1, glycidol is successfully converted into the desired product using both Lewis and Brønsted heterogeneous acid catalysts. The best results in terms of conversion and selectivity to solketal are obtained in the presence of supported metal triflates (see entries 3–5) and Nafion NR50 (see entry 1). No reaction took place using Montmorillonite K10, sulfonated activated charcoal (AC-SO$_3$H) and sulfonated mesoporous silica (MS-SO$_3$H) (see Table S1) due to their known lower total acidity (0.21 mmol/g for Montmorillonite K10, 0.15 mmol/g for MPS-SO$_3$H and 0.18 mmol/g for AC-SO$_3$H) [18]. Amberlyst-15 (sulfonated styrene-divinyl benzene resin with a total acidity of 4.7 mmol/g) promotes the quantitative conversion of glycidol but a dramatic reduction of the selectivity is observed due to a competitive glycidol oligomerization, as reported in our previous works for glycidol etherification with alcohols [18,22]. Among all the tested catalysts we evaluated the recyclability of Nafion NR50 and supported metal triflates in order to find the best catalytic system. Herein, in the presence of supported metal triflates no reaction took place due to active Lewis acid sites leaching. This phenomenon has been also recently reported in literature for Al(OTf)$_3$ on mesoporous silica-based catalysts during glycerol acetalization [23]. On the contrary, Nafion NR50 retains its high activity both in terms of conversion (90%) and selectivity to solketal (85%). It is worth to mention that the reported synthetic approach occurs with a 100% of atom economy with no formation of water. This aspect is crucial to avoid the undesirable deactivation of the sulfonic sites on Nafion NR50. The high activity of Nafion NR50 is related to Bronsted acidic sites and its perfluorinated polymeric structure as below confirmed by DFT calculations.

Table 1. Glycidol conversion to solketal in the presence of heterogeneous catalysts.

Experiment	Catalyst	Conversion (%)	Selectivity to Solketal (%)	Yield (%)
1	Nafion NR50	90	88	79
2	Amberlyst-15	100	8	8
3	Bi(OTf)$_3$ on MS	100	86	86
4	Al(OTf)$_3$ on MS	100	93	93
5	Fe(OTf)$_3$ on MS	100	87	87
6	No catalyst	0	-	-

Reaction conditions: glycidol/acetone moles ratio 1:43, t = 24 h, reflux, catalyst loading 10 wt %; MS: mesoporous silica.

With the best catalytic system, Nafion NR 50, we evaluated the effect of the temperature performing the reaction at room temperature. Herein, we observed only 24% conversion and 50% selectivity to solketal after 24 h due to the competitive glycidol oligomerization. As for the catalyst loading, reducing it from 10 wt % to 5 wt % only 58% of conversion with a total selectivity to solketal are achieved after 24 h. However, increasing the catalyst loading to 20 wt % allows to speed up the reaction and reach total conversion and selectivity to solketal. To evaluate the best acetone/glycidol ratio, catalytic runs were performed using a ratio of 20:1 under reflux for 18 h in the presence of 20 wt % of Nafion NR50. Results show a decrease of selectivity to solketal (80%) owing to glycidol oligomerization due to the more concentrated environment. Therefore, we continued our study by using the optimal conditions (acetone/glycidol molar ratio of 43, 20 wt % of Nafion NR50 and reflux conditions). The effect of the reaction time on conversion and selectivity under these optimized reaction conditions is shown in Figure 1. Nafion NR50 promotes the quantitative conversion (99%) of glycidol to solketal in 18 h with total selectivity to the desired product with a calculated TOF of 20 h^{-1}.

Moreover, the catalyst is stable under these reaction conditions and retains high efficiency during four consecutive cycles (see Figure S1 in Supplementary Materials). The recycled acetone has been characterized by GC-FID and analyses have demonstrated the high purity ensuring its potential reuse. This aspect is crucial at industrial level where the possibility to recycle the solvents increases the sustainability of the whole process with a drastic reduction of costs and environmental impacts [24].

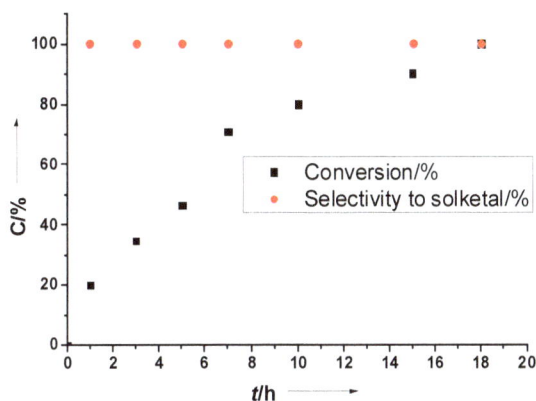

Figure 1. Glycidol conversion to solketal using Nafion NR50 (reaction conditions: glycidol/acetone moles ratio 1:43, catalyst loading 20 wt %, reflux).

Finally, in order to verify the generality of the studied reaction, we extended the substrate scope by using different ketones under the optimized reaction conditions (reflux, glycidol/ketone in moles ratio of 43, 18 h, 20 wt % of Nafion NR50). In details, methylethylketone (MEK) and 2-pentanone have been selected since the corresponding acetals can be opportunely used as building blocks to prepare high-value products such as monoalkyl glyceryl ethers [11]. Results are reported in Scheme 1. Glycidol is favorably converted into the corresponding acetals in both cases with high yields and selectivities, and glycidol oligomers are observed as by-products.

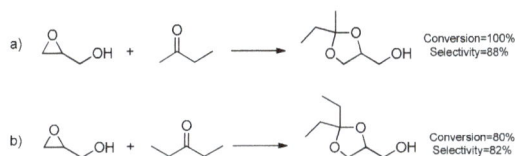

Scheme 1. Glycidol ketalization with ketones: methylethylketone (**a**), and 2-pentanone (**b**).

2.2. Theoretical Investigation of the Reaction Mechanism

The mechanism of the reaction between glycidol and acetone catalyzed by the best performing Nafion NR50 has been investigated by DFT calculations.

Two possible mechanistic scenarios have been investigated, see Figure 2. Pathway 1 implies the coordination of the epoxide to the catalyst trough formation of a hydrogen bond with the nucleophilic attack of the ketone that opens the ring in the following step; in pathway 2, instead, the catalyst activates acetone towards the nucleophilic addition of the -OH group on the glycidol and the epoxy ring opening occurs in the last step. The oxygen atom of the ketone moiety seems to be more nucleophile then the epoxy one, as proved by the almost 3 kcal/mol in favor of B′ adduct respect to B. Along pathway 1, after B formation, the ketone adds to the primary carbon atom of the epoxide with a concerted opening of the ring helped by the hydrogen transfer. The free energy barrier required is almost 24 kcal/mol. In C, the catalyst interacts with the substrate by formation of an oxygen-carbon bond that stabilizes the intermediate. The competing pathway with the ketone adding to the secondary carbon atom of the epoxide is both kinetically and thermodynamically unfavored by 1.4 and 5.8 kcal/mol, respectively. From C, the ex-epoxy oxygen adds to the ex-carbonyl carbon with the hydrogen returning to the sulfonate group on the catalyst and a barrier of almost 11.5 kcal/mol. Moving to pathway 2, the hydroxyl group of glycidol adds to the carbonyl carbon of acetone with a barrier of almost 21 kcal/mol.

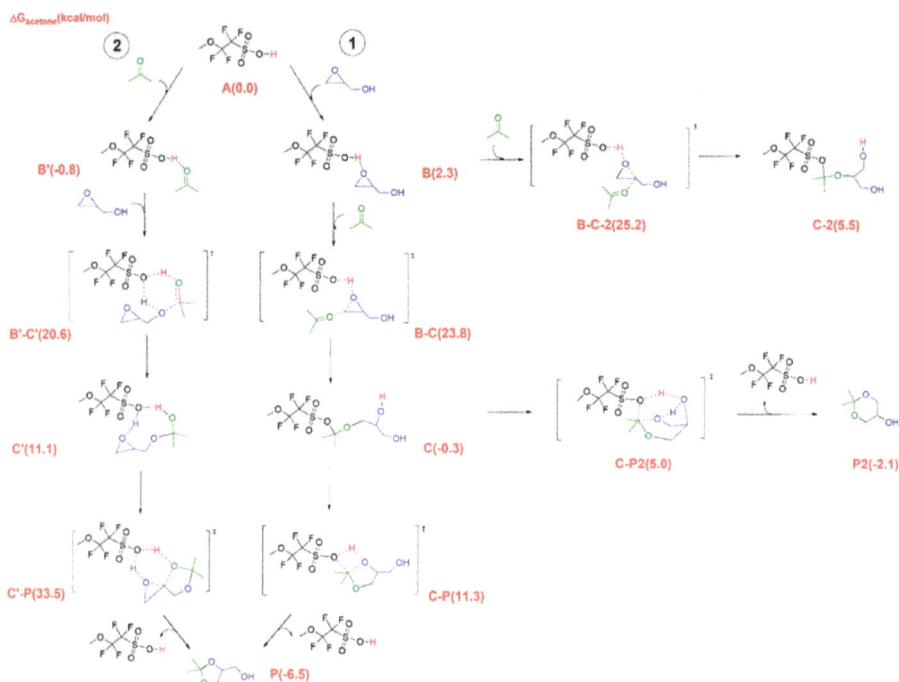

Figure 2. Mechanistic pathways investigated and corresponding free energies (kcal/mol in acetone).

The formation of the oxygen-carbon bond occurs simultaneously with two hydrogen transfers: the hydrogen of the sulphonic group transfers on the substrate and the hydrogen of the hydroxyl moiety transfers on the catalyst forming the intermediate C′ almost 10 kcal/mol lower than B′-C′. The following transition state C′-P consists in the opening of the epoxide ring concerted with the closure of the five member ring of the product P, 6.5 kcal/mol more stable than the starting reactants. This last step requires almost 34 kcal/mol ruling out this pathway, nevertheless the initial step is more favored along pathway 2 then along pathway 1. In conclusion, the favored mechanism consists in the ring opening of the epoxide by the nucleophilic attack of the ketone as rate determining step followed by an easier closure of the five member ring leading to the product. The formation of the corresponding six term ring product, P2 in Figure 2, is kinetically favored by almost 6 kcal/mol thanks to a most favorable geometry of the ring closing transition state. However, P2 is almost 5 kcal/mol less stable than P, that represents the thermodynamic product of the reaction, in agreement with the experimental findings (see Supplementary Materials). In order to rationalize the great performances showed by Nafion NR50 respect to the other heterogeneous systems tested, we have performed additional calculations for the reaction occurring in presence of sulfonated silica, as example of a not active catalyst for the considered reaction.

The favored pathway 1 of Figure 2 has been calculated for the catalytic system showed in Scheme S2 (see Supplementary Materials). The rate determining barrier for the nucleophilic attack of acetone to glycidol (B-C in in Figure S4) with the concerted ring opening assisted by the sulfonate moiety on the catalyst requires 28.1 kcal/mol for the silica system, i.e., almost 4.5 kcal/mol more than for the Nafion model. This result allows us to conclude that the great catalytic activity of Nafion is ascribed not only to the known higher concentration of acid groups on the catalyst respect to silicas for example, but also to the higher acidity of these groups that result to be more able to activate the glycidol towards the nucleophilic attack of the ketone, increasing meaningfully the yields of solketal formation.

3. Materials and Methods

3.1. Materials

Glycidol 96%, acetone, 3-pentanone, 2-butanone, Nafion NR 50 (0.7 mmol/g), Montomorillonite K10, activated charcoal, cetyltrimethylammonium bromide (CTAB), tetraethyl orthosilicate (TEOS), Bi(OTf)$_3$, Al(OTf)$_3$, Fe(OTf)$_3$, Amberlyst-15 (acidity 4.7 mmol/g) and sulfuric acid were purchased from Sigma-Aldrich. Glycidol and acetone were distilled before experiments. Mesoporous silica (MPS), MPS-supported metal triflates (Al(OTf)$_3$, Bi(OTf)$_3$ and Fe(OTf)$_3$)), sulfonated activated charcoal and sulfonated MPS were synthesized and characterized as described in our previous publication [18].

3.2. Catalytic Conversion of Glycidol to Solketal: General Conditions

In these experiments, 350 μL of glycidol and 15.0 mL of acetone (1:43 moles ratio) were mixed together in a round bottom flask under magnetic stirring (300 rpm) for 24 h under reflux conditions in the presence of an appropriate amount of heterogeneous catalyst (glycidol/catalyst weight ratio of 10). Afterwards, heterogenous catalyst was removed by filtration, acetone was removed using a rotary evaporator and the reaction products were analysed by GC-FID.

3.3. Gas-Chromatographic (GC-FID) Analyses

GC-FID analyses were carried out by using a Thermo Trace GC equipped with a Famewax polar column (30 m × 0.32 mm i.d.). The initial oven temperature was 40 °C, then programmed to heat to 160 °C at 5 °C min^{-1}, then to 240 °C at 20 °C min^{-1} and held at 240 °C for 5 min with a flow rate of 1.0 mLmin^{-1} (splitless injection mode was used). The injection volume was 1 μL. The FID temperature was 280 °C and 230 °C for the inlet. The integrated areas were converted into mole percentages for each component present in the sample by using calibration curves prepared for all the components and 3-ethoxy-1,2-propanediol as internal standard. The data obtained were used to calculate the conversion and selectivity of the reactant species. Conversion (C) and selectivity (S) to products were calculated as follows:

$$\text{Glycidol conversion } (\%) = \frac{(initial\ mol\ of\ glycidol - final\ mol\ of\ glycidol)}{initial\ mol\ of\ glycidol} * 100 \qquad (1)$$

$$\text{Selectivity } (\%) = \frac{mol\ of\ defined\ product}{mol\ of\ reacted\ glycidol} * 100 \qquad (2)$$

$$\text{Yield } (\%) = [\text{conversion } (\%) * \text{selectivity } (\%)]/100 \qquad (3)$$

The relative standard deviation of three replicates is lower than 4% in all cases.

4. Conclusions

In conclusion, we have reported the selective preparation of solketal through glycidol (obtained as value-added product from Epicerol process) acetalization with acetone in the presence of Nafion as heterogeneous catalyst. Notably, using a low catalyst loading of Nafion (1.5% in moles as SO$_3$H group toward glycidol) we demonstrated the quantitative conversion of glycidol to the desired product in 18 h of reaction under mild conditions (reflux, acetone/glycidol molar ratio of 43). Nafion is also stable allowing to be reused for several reaction cycles without any loss of activity and selectivity. The study has been also extended to other ketones and solketal derivates are produced under the optimized reaction conditions with good yields and selectivity. The use of a heterogeneous catalyst to perform this reaction represents the innovative part of this research together with the theoretical investigation of the reaction mechanism. In fact, the calculations performed allowed to discriminate the energetically favored mechanistic pathway, highlighting that the opening of the glycidol ring is likely to occur in the first step of the reaction, concerted with the nucleophilic attack of acetone to the epoxy carbon. In fact, the alternative mechanism that sees the three member ring opening in the last step concerted with the solketal five member ring closure is almost 10 kcal/mol more energy requiring. Finally, the mechanistic pathway calculated for the system simulating the MS-SO$_3$H catalyst showed that the fluorinated polymeric skeleton of Nafion is more able to activate glycidol towards acetone addition decreasing the decisive reaction barrier.

Supplementary Materials: The following are available online at http://www.mdpi.com/2073-4344/8/9/391/s1. Figure S1. Nafion NR50 recyclability. Figure S2. 1H-NMR (CDCl3, 400 MHz) spectrum of reaction mixture. Figure S3. 13C-NMR (CDCl3, 100 MHz) spectrum of reaction mixture. Figure S4. Mechanistic pathways investigated and corresponding free energies (kcal/mol in acetone) for the sulfonated-silica catalyzed reaction. Table S1. Glycidol conversion to solketal. Scheme S1. Nafion NR 50 modeled structure. Scheme S2. Sulfonated-silica modelled structure.

Author Contributions: M.R., L.F. and R.C. performed the experiments; R.C. and L.F. wrote the paper; F.C. and A.P. conceived and designed the experiments and discussed the results; T.T. supported the analysis of data and discussed the results.

Funding: This research was funded by University of Salerno, ORSA167988.

Acknowledgments: This work was financially supported by research fund "FARB 2016", University of Salerno (ORSA167988).

Conflicts of Interest: The authors declare no conflict of interest.

References

1. Cespi, D.; Passarini, F.; Vassura, I.; Cavani, F. Butadiene from biomass, a life cycle perspective to address sustainability in the chemical industry. *Green Chem.* **2016**, *18*, 1625–1638. [CrossRef]
2. Tripodi, A.; Bahadori, E.; Cespi, D.; Passarini, F.; Cavani, F.; Tabanelli, T.; Rossetti, I. Acetonitrile from Bioethanol Ammoxidation: Process Design from the Grass-Roots and Life Cycle Analysis. *ACS Sustain. Chem. Eng.* **2018**, *6*, 5441–5451. [CrossRef]
3. Anastas, P.T.; Warner, J.C. Principles of green chemistry. In *Green Chemistry: Theory and Practice*; Oxford University Press: Oxford, UK, 1998; pp. 29–56, ISBN 0-19-850234-6.
4. Cespi, D.; Cucciniello, R.; Ricciardi, M.; Capacchione, C.; Vassura, I.; Passarini, F.; Proto, A. A simplified early stage assessment of process intensification: Glycidol as a value-added product from epichlorohydrin industry wastes. *Green Chem.* **2016**, *18*, 4559–4570. [CrossRef]
5. Pagliaro, M.; Rossi, M. *The Future of Glycerol*; RSC Publishing: Cambridge, UK, 2010; pp. 1–25, ISBN 978-1-84973-046-4.
6. Canale, V.; Tonucci, L.; Bressan, M.; d'Alessandro, N. Deoxydehydration of glycerol to allyl alcohol catalyzed by rhenium derivatives. *Catal. Sci. Technol.* **2014**, *4*, 3697–3704. [CrossRef]
7. Zhou, C.-H.; Beltramini, J.N.; Fan, Y.-X.; Lu, G.Q. Chemoselective catalytic conversion of glycerol as a biorenewable source to valuable commodity chemicals. *Chem. Soc. Rev.* **2008**, *37*, 527–549. [CrossRef] [PubMed]
8. Cucciniello, R.; Pironti, C.; Capacchione, C.; Proto, A.; Di Serio, M. Efficient and selective conversion of glycidol to 1, 2-propanediol over Pd/C catalyst. *Catal. Commun.* **2016**, *77*, 98–102. [CrossRef]
9. Cucciniello, R.; Ricciardi, M.; Vitiello, R.; Di Serio, M.; Proto, A.; Capacchione, C. Synthesis of Monoalkyl Glyceryl Ethers by Ring Opening of Glycidol with Alcohols in the Presence of Lewis Acids. *ChemSusChem* **2016**, *9*, 3272–3275. [CrossRef] [PubMed]
10. Vicente, G.; Melero, J.A.; Morales, G.; Paniagua, M.; Martín, E. Acetalisation of bio-glycerol with acetone to produce solketal over sulfonic mesostructured silicas. *Green Chem.* **2010**, *12*, 899–907. [CrossRef]
11. Samoilov, V.O.; Onishchenko, M.O.; Ramazanov, D.N.; Maximov, A.L. Glycerol Isopropyl Ethers: Direct Synthesis from Alcohols and Synthesis by the Reduction of Solketal. *ChemCatChem* **2017**, *9*, 2839–2849. [CrossRef]
12. Nanda, M.R.; Zhang, Y.; Yuan, Z.; Qin, W.; Ghaziaskar, H.S.; Xu, C. Catalytic conversion of glycerol for sustainable production of solketal as a fuel additive: A review. *Renew. Sustain. Energy Rev.* **2016**, *56*, 1022–1031. [CrossRef]
13. Li, L.; Korányi, T.I.; Sels, B.F.; Pescarmona, P.P. Highly-efficient conversion of glycerol to solketal over heterogeneous Lewis acid catalysts. *Green Chem.* **2012**, *14*, 1611–1619. [CrossRef]

14. Iranpoor, N.; Kazemi, F. Ru(III) catalyses the conversion of epoxides to 1,3-dioxolanes. *Synth. Commun.* **1998**, *28*, 3189–3193. [CrossRef]

15. Iranpoor, N.; Adibi, H. Iron(III) Trifluoroacetate as an Efficient Catalyst for Solvolytic and Nonsolvolytic Nucleophilic Ring Opening of Epoxides. *BCSJ* **2000**, *73*, 675–680. [CrossRef]

16. Procopio, A.; Dalpozzo, R.; De Nino, A.; Maiuolo, L.; Nardi, M.; Russo, B. Synthesis of Acetonides from Epoxides Catalyzed by Erbium(III) Triflate. *Adv. Synth. Catal.* **2005**, *347*, 1447–1450. [CrossRef]

17. Ricciardi, M.; Passarini, F.; Vassura, I.; Proto, A.; Capacchione, C.; Cucciniello, R.; Cespi, D. Glycidol, a Valuable Substrate for the Synthesis of Monoalkyl Glyceryl Ethers: A Simplified Life Cycle Approach. *ChemSusChem* **2017**, *10*, 2291–2300. [CrossRef] [PubMed]

18. Ricciardi, M.; Passarini, F.; Capacchione, C.; Proto, A.; Barrault, J.; Cucciniello, R.; Cespi, D. First Attempt of Glycidol-to-Monoalkyl Glyceryl Ethers Conversion by Acid Heterogeneous Catalysis: Synthesis and Simplified Sustainability Assessment. *ChemSusChem* **2018**, *11*, 1829–1837. [CrossRef] [PubMed]

19. Della Monica, F.; Buonerba, A.; Grassi, A.; Capacchione, C.; Milione, S. Glycidol: An Hydroxyl-Containing Epoxide Playing the Double Role of Substrate and Catalyst for CO₂ Cycloaddition Reactions. *ChemSusChem* **2016**, *9*, 3457–3464. [CrossRef] [PubMed]

20. Mohammadpoor-Baltork, I.; Khosropour, A.R.; Aliyan, H. Efficient Conversion of Epoxides to 1,3-Dioxolanes Catalyzed by Bismuth(III) Salts. *Synth. Commun.* **2001**, *31*, 3411–3416. [CrossRef]

21. Lange, J.-P. Don't Forget Product Recovery in Catalysis Research—Check the Distillation Resistance. *ChemSusChem* **2017**, *10*, 245–252. [CrossRef] [PubMed]

22. Ricciardi, M.; Cespi, D.; Celentano, M.; Genga, A.; Malitesta, C.; Proto, A.; Capacchione, C.; Cucciniello, R. Bio-propylene glycol as value-added product from Epicerol® process. *Sustain. Chem. Pharm.* **2017**, *6*, 10–13. [CrossRef]

23. Tayade, K.N.; Mishra, M.; Munusamy, K.; Somani, R.S. Synthesis of aluminium triflate-grafted MCM-41 as a water-tolerant acid catalyst for the ketalization of glycerol with acetone. *Catal. Sci. Technol.* **2015**, *5*, 2427–2440. [CrossRef]

24. Cucciniello, R.; Cespi, D. Recycling within the Chemical Industry: The Circular Economy Era. *Recycling* **2018**, *3*, 22. [CrossRef]

catalysts

MDPI

Review

Mixed-Oxide Catalysts with Spinel Structure for the Valorization of Biomass: The Chemical-Loop Reforming of Bioethanol

Olena Vozniuk [1,2], Tommaso Tabanelli [1], Nathalie Tanchoux [2], Jean-Marc M. Millet [3], Stefania Albonetti [1], Francesco Di Renzo [2] and Fabrizio Cavani [1,*]

[1] Department of Industrial Chemistry "Toso Montanari", Alma Mater Studiorum Università di Bologna, 40136 Bologna, Italy; vozniuk89@gmail.com (O.V.); tommaso.tabanelli@unibo.it (T.T.); stefania.albonetti@unibo.it (S.A.)

[2] Institut Charles Gerhardt, UMR 5253 UM-CNRS-ENSCM, 34296 Montpellier, France; nathalie.tanchoux@enscm.fr (N.T.); Francesco.Di-Renzo@enscm.fr (F.D.R.)

[3] Univ Lyon, Université Claude Bernard Lyon 1, CNRS, IRCELYON–UMR 5256, 69626 Villeurbanne, France; jean-marc.millet@ircelyon.univ-lyon1.fr

* Correspondence: fabrizio.cavani@unibo.it; Tel.: +39-0592-093-680

Received: 29 July 2018; Accepted: 12 August 2018; Published: 14 August 2018

Abstract: This short review reports on spinel-type mixed oxides as catalysts for the transformation of biomass-derived building blocks into chemicals and fuel additives. After an overview of the various methods reported in the literature for the synthesis of mixed oxides with spinel structure, the use of this class of materials for the chemical-loop reforming of bioalcohols is reviewed in detail. This reaction is aimed at the production of H_2 with intrinsic separation of C-containing products, but also is a very versatile tool for investigating the solid-state chemistry of spinels.

Keywords: spinels; metal ferrites; bioethanol; chemical-loop reforming; syngas

1. Introduction

Spinel oxides with the general formula AB_2O_4 are chemically and thermally stable materials suitable for several applications, including catalysis.

The ideal stoichiometric spinel structure is assumed by oxides with average cation charge of 2.33, like, for instance, magnetite Fe_3O_4, with one divalent and two trivalent Fe cations. The presence of cations with different charges is at the basis of most catalytic properties of spinels, allowing internal redox reactions which make easier reduction–reoxidation cycles of the catalyst. In the spinel structure, the oxygen anions are distributed in an approximate cubic close-packing and the cations are distributed in the interstices between the oxygen anions. Only a fraction of the interstices are occupied by cations, namely eight tetrahedral interstices (A sites) and 16 octahedral interstices (B sites) in a cell containing 32 oxygen anions, each anion being located at the corner between two octahedra and one tetrahedron. The occupied cation sites form rows of octahedra joined edge-to-edge and connected by tetrahedra (Figure 1).

The distribution of different cations in the A and B sites essentially depends on their crystal field stabilisation in coordination four or six. The effect of this distribution on catalytic properties is not negligible. When the spinel structure is interrupted by a surface, octahedral B sites are more exposed than tetrahedral A sites. As a consequence, B sites have been considered to represent the most effective catalytic sites [1]. However, this assumption has somehow to be qualified, as the internal charge transfers needed to close a catalytic cycle imply both site B and site A, and both sites are involved in oxygen mobility. Moreover, the reduction of the surface implies a severe reorganisation which significantly modifies the distribution of the sites [2]. Reduction of the oxide can reach complete destruction of the spinel phase when the material is used as oxygen carrier in a chemical-loop cycle.

The spinel structure is tolerant of significant deviations from the average cation valence 2.33, with significant effects on catalytic activity. The incorporation of cations with higher positive charge is possible if accompanied by the formation of cation vacancies, which allow the charge balance of the solid to be kept. A typical example is γ-Fe$_2$O$_3$ (maghemite), a cation-defective spinel. The absence of divalent Fe^{2+} cations in maghemite has been suggested to explain the lower effectiveness of maghemite as Fenton catalyst by comparison with the isostructural not-defective magnetite Fe$_3$O$_4$ [3]. On the contrary, partial oxidation of spinels has been reported to create oxygen defects, on which reactive ·OH radicals are easily formed [4].

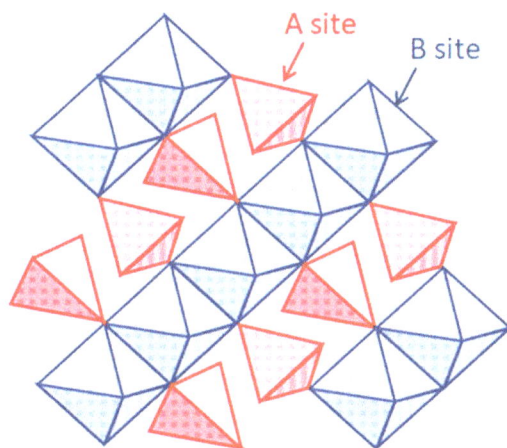

Figure 1. Spinel structure. Cations occupy the centre of tetrahedra (sites A) and octahedra (sites B). Oxygen anions are at the corner between two octahedra and one tetrahedron.

Overall, versatile catalytic properties are dependent on the chemical composition and nature of substituted ions, charges and their distribution among the octahedral (Oh) and tetrahedral (Td) sites, which in turn are affected by the synthesis method used [5–8]. Indeed, many methods have been developed to synthesize spinel oxides, such as, amongst others, solid-state synthesis (mechanical milling/alloying), and wet chemical routes: sol–gel, coprecipitation, reverse micelles, hydrothermal/solvothermal synthesis, electrochemical synthesis, and others (Table 1).

Table 1. Methods reported in the literature for spinel synthesis.

Preparation Method	Reference
Coprecipitation	[9–23]
Sol–gel	[24–46]
Hydrothermal	[47–55]
Solvothermal	[56–60]
Microemulsion/Reverse micelles	[61–68]
Template	[66–72]
Mechanical milling	[73–81]
Plasma	[82,83]
Flux growth	[84–86]
Solid phase	[87]
Combustion	[88–90]
Microwave combustion	[91–93]
Microwave hydrothermal	[94–96]
Pechini method	[97–100]

Table 1. *Cont.*

Preparation Method	Reference
Electrochemical	[101]
Electrospinning	[102]
Thermal treatment	[103,104]
Ultrasonic wave-assisted ball milling	[105]
Spray pyrolysis	[106]
Aerosol	[107]
Forced hydrolysis	[108]
Glycol-thermal	[109]
Refluxing	[110]

The choice of the preparation method is generally driven by the stability of the specific spinel composition targeted and by the requirement of specific textural, chemical or magnetic properties of the final product. Obtaining particles smaller than 10 nm is usually possible by proper tuning of most low-temperature methods. The temperature of post-treatment is the most important factor affecting the size of the spinel particles. The formation of small particles at high temperature is possible by reducing the time spent at the synthesis temperature, for instance in methods of solution combustion or spray pyrolysis. The stability of these small particles in the temperature and redox conditions of the catalytic application is, however, often questionable. The morphology of the spinel particles has been controlled by combination of pH, temperature and flow conditions in several methods.

The preparation methods largely differ by their economics, their energy requirement and their environmental impact. Classical precipitation methods can be hampered by the need of operating in the high-pH field in which all the concerned cations are out of their solubility domain, with consequent rejection of alkaline wastewaters. Methods aimed at a more precise control of particle size by confinement of the precursors, as the emulsion or template methods, imply the use of relatively costly organic additives. This is also the drawback of methods such as the sol–gel, Pechini or alginate methods, in which the homogeneous dispersion of the precursor cations in a matrix favours the formation of spinels with compositions difficult to form by coprecipitation. Some energy-intensive methods, like flux growth, are not intended to form the high-surface-area materials preferred for catalysis but aim for the formation of materials with specific solid-state properties.

Several uses of spinel-type mixed oxides as catalysts for a variety of reactions have been reported; Table 2 summarizes the main ones.

Table 2. Applications of spinel mixed-metal oxides as catalysts.

Reaction	References
Oxidative cleavage of styrene to benzaldehyde with H_2O_2	[5,49]
Oxidation of cyclohexane to cyclohexanol/cyclohexanone with O_2 or H_2O_2	[6,7]
Hydroxylation of benzene/phenol to phenol/diphenols with H_2O_2	[111]
Oxidation of vanyllol to vanillin with air	[112]
Oxidation of benzyl alcohol to benzaldehyde with H_2O_2	[113]
Oxidation of monoterpenic alkenes with O_2	[114]
Oxidation of 5-hydroxymethylfurfural to 2,5-furandicarboxylic acid (hmf to fdca) with H_2O_2 or O_2	[115–117]
Oxidation of aniline to azoxybenzene with H_2O_2	[118]
Oxidation of toluene to benzaldehyde with H_2O_2	[119]
Oxidation of ethanol to acetaldehyde with O_2	[120]
Oxidation of veratryl alcohol to veratryl aldehyde with O_2	[121]
Ketonisation of butanol to heptanone	[122]
Total oxidation of voc with air	[123]
Friedel–crafts acylation	[124]
Knoevenagel condensation	[125]
Reduction of ketones	[126]
Reduction of nitroarenes	[127]
Methylation (alkylation) of phenolics, aniline, pyridine	[128]
Methanol, ethanol reforming (by means of chemical-loop)	[129–137]

It is shown that most applications investigated are for the oxidation of organic substrates, for example, for the synthesis of aldehydes or acids. In the field of biomass valorization, worthy of note are the recent papers on the oxidation of 5-hydroxymethylfurfural (HMF) to 2,5-furandicarboxylic acid (FDCA) with O_2, and catalytic transformation involving bioalcohols, mainly bioethanol. Recently, Jain et al. reported on the use of spinel catalysts with composition $Li_2CoMn_3O_8$ as efficient catalysts for the selective oxidation of HMF to FDCA with 80% isolated yield in a gram-scale reaction [116]. The nanocrystalline spinel was synthesized by a gel pyrolysis method using urea and citric acid as complexing agent. Prompted by the activities of Co- and Mn-based homogeneous catalyst systems such as $Co(OAc)_2/Mn(OAc)_2/HBr$ used in HMF oxidation, spinel $MnCo_2O_4$-supported Ru nanoparticles were synthesized and applied as heterogeneous nanocatalysts for HMF oxidation by Mishra et al. [117], under base-free conditions. An important role was ascribed to the acidic sites on the spinel surface in affording 99% yield to FDCA.

2. Spinels as Catalysts for the Chemical-Loop Reforming (CLR) of Bioethanol

The reforming of bioethanol, and in general of bioalcohols, to syngas, has been the object of several investigations during the latest years. This reaction can be carried out in the chemical-loop mode, that is, by alternating the bioalcohol and steam over an O-carrier [129–137], the so-called chemical-loop reforming (CLR). Amongst the most promising materials, ferrospinels offer the advantage of a wide flexibility of composition, structural stability and tunable redox properties. On the other hand, the choice of ethanol as reducing agent has also several advantages: its renewable origin, availability in large quantities at low cost, together with the possibility to decompose at a relatively lower temperature with formation of a hydrogen-rich mixture. CLR is aimed at the production of "clean H_2" with an inherent CO_x separation. The main principle of the CLR process is that an oxygen-storage material is first reduced by ethanol stream (T = 400–500 °C), and then reoxidized by water (T = 300–450 °C) to produce hydrogen and to restore the original oxidation state of the looping material (Figure 2).

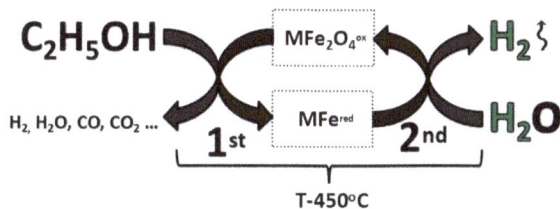

Figure 2. The chemical-loop reforming of ethanol over modified ferrospinels.

Different M-modified MFe_2O_4 spinel-type mixed oxides were synthesized and tested as ionic oxygen and electrons carriers to generate hydrogen by water reduction, after a reductive step of the oxides carried out with ethanol. The aim was to develop materials showing the structural stability needed to undergo complete reversible redox cycling upon chemical looping. Spinels containing Co, Mn, Cu or Cu/Co, Cu/Mn, Co/Mn and alkaline earth metals Ca or Mg as divalent cations were prepared, characterised and tested. The nature of the cations affected the reactivity of the spinels, in regard to both the nature of the products formed during ethanol oxidation along with the purity of the hydrogen produced during the water-reduction step.

Regarding the behaviour of bare Fe_3O_4, during the reduction step it formed Fe^0 which then was converted into Fe_3C (cementite). However, the formed carbide decomposes into metallic iron and carbon ($Fe_3C{\rightarrow}3Fe^0{+}C$) and in consequence catalyses the growth of graphitic filaments. In order to reduce the formation of coke, a short reduction time of 5 min only was used, since the formation of cementite was slightly delayed at the beginning of the reduction step. The reducibility of magnetite was improved by incorporation of several transition metals like Co, Cu and Ni into the spinel structure.

The structure of MFe_2O_4 ferrospinels, prepared via a coprecipitation route, was identified by means of XRD. A broadening of the diffraction patterns was observed for Mn ferrospinels, that is, $MnFe_2O_4$, $Cu_{0.5}Mn_{0.5}Fe_2O_4$ and $Co_{0.5}Mn_{0.5}Fe_2O_4$, attributed to a decrease in particle size. Table 3 compiles the specific surface area (SSA), the crystallite size (calculated by Scherrer equation) and the particle size of the fresh powders calcined at 450 °C for 8 h.

Table 3. Specific surface area (SSA) and crystallite/particle size of spinels with various compositions.

Sample Name	SSA, m^2/g	Crystallite Size, nm	Particle Size (d_{BET}), nm
$CuFe_2O_4$	60	6.9	18.3
$Cu_{0.5}Co_{0.5}Fe_2O_4$	67	10.4	16.5
$CoFe_2O_4$	69	12	16.2
$Co_{0.5}Mn_{0.5}Fe_2O_4$	141	3.5	8
$Cu_{0.5}Mn_{0.5}Fe_2O_4$	112	-	10
$MnFe_2O_4$	165	-	6.9

Temperature-programmed reduction (TPR) was used to characterise the redox properties of samples (Figure 3). The reduction of Fe strongly depends on the presence of another metal in the ferrite. Two main steps of the reduction were shown: (i) the reduction of iron oxide to metallic iron and (ii) the reduction of the incorporated metal oxide to its corresponding metal or sub-oxide. Despite the overlapping of the two steps, based on the nature of the foreign metals and their reduction potentials, combined with the experimental amount of consumed H_2, it was possible to draw the reduction scheme shown in Figure 4.

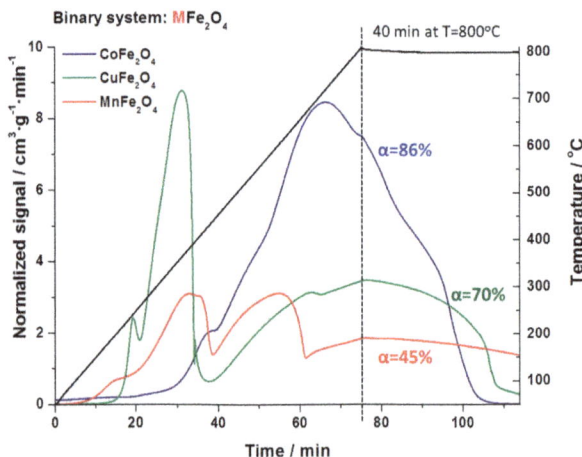

Figure 3. TPR profile of MFe_2O_4 ferrospinels, where M = Cu, Co or Mn; α = degree of reduction (%).

As an example, in the TPR profile of $CuFe_2O_4$, a peak at 240 °C can be attributed to the reduction of the copper oxide to the metallic copper, whereas a second peak at ~340 °C can be attributed to a primary stepwise reduction of the spinel with a final formation of Cu^0 and Fe_3O_4; the further reduction of Fe_3O_4 to FeO and of the latter to Fe^0 appears at higher temperatures. In the case of $CoFe_2O_4$, primary spinel reduction started to occur at about 400 °C, with a stepwise formation of CoO and Fe_3O_4. With the Mn-containing spinel, the reduction of MnO to metallic Mn turned out to be very difficult, due to its highly negative reduction potential (−1.18 eV), and thus this step occurs only at very high temperatures. In fact, the total reduction extent of $CoFe_2O_4$ (α = 86%) was much higher than that of $MnFe_2O_4$ (α = 45%) samples, which can be explained by the formation of the hardly reducible MnO or of Mn_xFe_yO oxide.

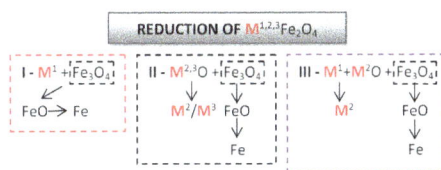

Figure 4. Reduction scheme of MFe_2O_4 ferrospinels.

The reactivity of spinels has been tested in the CLR of bioethanol; the role of the first step, that is, the reduction of the spinel with the alcohol at 450 °C, was aimed at maximizing the spinel reduction extent along with minimizing deactivation, an effect of coke accumulation. In other words, the reduction degree should be monitored as closely as possible, while both limiting the formation of coke, and maintaining the reoxidizability of the material during the second step to regenerate the starting spinel. This condition was essential, in order to allow the looping material to repeat the cycle as many times as possible. Another important point is that during the reduction of the spinel with ethanol, the latter is not only decomposed to light gases (that is, CO, CO_2, CH_4 and H_2), but also is oxidised to several compounds, ranging from C2 (acetaldehyde, acetic acid), to C3 (acetone), C4 and higher homologues. These valuable compounds can be easily separated from the light gases and could contribute to the overall process's economic sustainability. The nature and amount of these "by-products" turned out to be a function of spinel composition, as well as of conditions used for the two steps of the chemical loop.

Figure 5 summarizes the integrated values for H_2 produced during the second step, that is, the reoxidation of the reduced MFe_2O_4 spinel (referred to as one complete cycle of 20 min for each one of the two steps) carried out with steam. For a better understanding of the results, some important values are provided in Table 4.

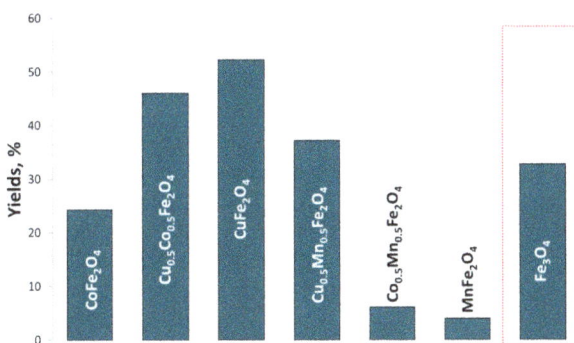

Figure 5. Integrated yields for H_2 produced during second reoxidation step carried out with steam at 450 °C and MFe_2O_4 ferrospinels (note: Listed data correspond to the values obtained after 1 complete cycle of 20 min).

The following aspects are worthy of being mentioned:

(a) Mn incorporation into Fe_3O_4 with generation of the corresponding ferrites showed its positive aspect on lowering the amount of coke that accumulated during the first step carried out with ethanol, see C %$_w$ (CHNS) in Table 4;

(b) Mn incorporation also increased the H_2/CO_x ratio, which follows from the previous statement. It is important to notice that the higher the H_2/CO_x ratio, the more 'pure' is the H_2 generated during the second step. For comparison, Fe_3O_4 itself accounts for $H_2/CO_x = 3.5$, whereas $MnFe_2O_4$ ($H_2/CO_x = 15$) and $Co_{0.5}Mn_{0.5}Fe_2O_4$ ($H_2/CO_x = 15$) showed much higher values;

(c) the incorporation of Cu (alone, or together with either Co or Mn) has a beneficial effect on the total amount of H_2 produced from H_2O, compared to Fe_3O_4. Hence, the best performance was shown by $CuFe_2O_4$ (Y-52%), $Cu_{0.5}Co_{0.5}Fe_2O_4$ (Y-46%) and $Cu_{0.5}Mn_{0.5}Fe_2O_4$ (Y-37%);

(d) incorporation of Cu/Co led to the increase of the n_{H2}/n_{Eth} ratio, as for $CuFe_2O_4$ ($n_{H2}/n_{Eth} = 1.2$), and $Cu_{0.5}Co_{0.5}Fe_2O_4$ ($n_{H2}/n_{Eth} = 1.0$). This can be correlated to the feasibility of producing H_2 starting from bioethanol being based on the H_2 vs ethanol heating values (referred to as LHV (Lower Heating Value), MJ/kg): 119, 96 and 28.86, respectively. In other words, the higher the n_{H2}/n_{Eth}, the higher is the potential efficiency of the CLR process. Of course, there are many more aspects that have to be undertaken in order to calculate the actual cost of the CLR process, and to estimate a final price of H_2 produced via CLR technology and compare it to the existing ones (not encompassed in this study).

Table 4. Chemical-loop process parameters calculated for MFe_2O_4 ferrites in bioethanol reforming (note: Listed data correspond to the values obtained after 1 complete cycle of 20 min).

Sample Name	C $\%_W$ after 20 min Red. with Ethanol	H_2/CO_X	Moles of H_2/Moles of Ethanol (n_{H2}/n_{Eth})
$CoFe_2O_4$	11.6	6	0.5
$Cu_{0.5}Co_{0.5}Fe_2O_4$	16.3	3	1.0
$CuFe_2O_4$	6.9	3	1.2
$Cu_{0.5}Mn_{0.5}Fe_2O_4$	6.1	3	0.8
$Co_{0.5}Mn_{0.5}Fe_2O_4$	1.5	15	0.1
$MnFe_2O_4$	1.7	15	0.09
Fe_3O_4	5.3	3.5	0.7

Table 5 summarizes the H_2 produced during three consecutive cycles of 20 min. The following statements can be made:

(a) Consecutive utilization of $CoFe_2O_4$, $CuFe_2O_4$ and $Cu_{0.5}Co_{0.5}Fe_2O_4$ ferrospinels as looping materials resulted in higher amounts of produced hydrogen (given in moles) which surpass the value obtained over the reference material—Fe_3O_4;

(b) increasing the total *tos* from 20 to 60 min (which accounts the total time for the reduction/reoxidation step) leads to the decreasing of H_2/CO_x ratio, which in its turn affects the final purity of the target gas—H_2. However, this problem can be overcome by implementation of a three-step CLR process with the third step being carried out with air;

(c) $CuFe_2O_4$ showed the higher n_{H2}/n_{Eth} ratio of 1 (referring to the total value for three consecutive cycles) which was in fact twice as high as that obtained with Fe_3O_4 ($n_{H2}/n_{Eth} = 0.5$);

(d) on the other hand, under different conditions (not shown here), $CoFe_2O_4$ underwent the greatest extent of reduction during the first step, while being reoxidizable back to the spinel during the second step, and was able to maintain it throughout several repeated cycles. However, it showed the greater amount of accumulated coke, which formed CO when put in contact with steam during the second step;

(e) coke formation remained an issue for M-modified ferrospinels, which means that avoiding completely carbon deposition and its further accumulation is not possible.

Table 5. Chemical-loop process parameters calculated for MFe_2O_4 ferrites in bioethanol reforming (note: Listed data correspond to the values obtained after 3 complete cycles of 20 min).

Sample Name	H_2/CO_X	Moles of H_2/Moles of Ethanol
$CoFe_2O_4$	5	0.7
$Cu_{0.5}Co_{0.5}Fe_2O_4$	3	0.9
$CuFe_2O_4$	3	1.0
Fe_3O_4	3	0.5

Velasquez Ochoa et al. [138] studied the reduction mechanism of M-modified (Ni, Co and Fe) spinel oxides, where ethanol was the reductant. It was concluded that the first step in ethanol anaerobic decomposition appears to be the same for all samples and corresponds to acetaldehyde formation via dehydrogenation of ethanol. Further reduction of the solid was strongly dependent on the nature of incorporated M (Ni, Co or Fe), viz. acetaldehyde can be either oxidized to acetates ($NiFe_2O_4$), decomposed to CO and CH_4 ($CoFe_2O_4$) or completely oxidized (Fe_3O_4). As said above, Mn incorporation significantly reduced the coke formation during the first reduction step, which was attributed to the formation of a thermodynamically stable and hardly reducible layer of Mn_xFe_yO solid solution. Moreover, it predominantly favored dehydrogenation and condensation reactions leading to the formation of acetaldehyde and acetone, whereas Co/Cu incorporation facilitated total/partial oxidation of ethanol giving rise to high yields of H_2, CO_x and H_2O.

Recent study on $CoFe_2O_4$ and $FeCo_2O_4$ as oxygen carrier materials was performed by Carraro et al. [129,130]. During the reduction step with ethanol, $FeCo_2O_4$ was reduced faster compared to $CoFe_2O_4$. However, its performance during the reoxidation step was quite poor due to an inefficient oxidation by water steam, which is able to oxidize only the outer shell of the nanoparticles, resulting in small H_2 yield. On the other hand, $CoFe_2O_4$ sample was a more efficient oxygen carrier, which enabled the production of a larger amount of H_2 due to the residual presence of a reducible wüstite, which can be consecutively reoxidized/reduced in further looping cycles.

3. Other Materials as O-Carriers for Hydrogen Production via CLR

A comprehensive review on different oxygen carrier materials for the hydrogen production via chemical-loop processes was recently published by Protasova et al. [139]. The review encompasses information on the different perovskites and Ni/Fe/Cu/Ce-based oxygen carrier materials. Perovskites showed good results for the partial oxidation of methane, while with Fe-based materials, promising results also have been obtained. Several research groups have been exploring modifications of simple iron oxide (Fe_3O_4 and Fe_2O_3) in order to prevent deactivation [140], to lower the operating temperature [141] and to increase the structural stability and reducibility [142,143], and to increase the reaction rate for oxidation and total efficiency of the process [144]. Several studies were dedicated to different metal additives to iron oxide [145,146]. In addition, ternary metal systems have also been considered in the search for a better synergetic effect [147,148]. Several research groups have investigated the effect of various M-additives on the stability and redox behavior of iron oxide for chemical hydrogen storage using Pd, Pt, Rh, Ru, Al, Ce, Ti, Zr [149] and Al, Cr, Zr, Ga, V, Mo [150]. It was found out that Pd, Pt, Rh and Ru additives have an effect on promoting the reduction and lowering the reoxidation temperature of iron oxide. At the same time, Al, Ce, Ti, Zr, Cr, Ga and V additives prevent deactivation and sintering of iron oxide during repeated redox cycles. Some recent studies on developing of the novel and efficient oxygen carrier materials for chemical-loop applications highlight the special interest in spinel oxides [151–160] which, first of all, were explained by their ability to form thermodynamically stable spinel oxides which allow one to reobtain the initial spinel phase upon cycling, and in turn, increase the stability of the looping material itself.

4. Conclusions

Spinel-type mixed-metal oxides are extremely versatile systems useful as catalysts for a variety of reactions. Because of their chemical–physical properties, they are now studied for redox reactions involving biomass-derived building blocks aimed at the production of either chemicals or fuel components. An example is the reforming of bioalcohols into CO_x/H_2; this reaction can be carried out in the chemical-loop mode, which in principle should allow the intrinsic separation of H_2 from CO_x. The reaction also allows the study of the chemical–physical properties of spinels in terms of redox properties. Depending on the spinel composition, it is possible to obtain H_2 along several cycles, but during the spinel reduction step (with the bioalcohol) unfortunately also coke forms, which brings about the formation of CO_x during the spinel reoxidation step; the latter, however, should be aimed at the production of H_2 only. Studies showed that a complete recovery of the initial cycling material was possible, although a slow accumulation of coke takes place (1.0 wt % after 20 cycles or 100 min). This problem could be solved by periodically adding a third step to burn this coke left over by air.

If the production of only H_2 in the second chemical looping step is not a crippling issue, the process can be envisaged to valorize bioethanol. In fact, quite interestingly, the reduction of the spinel with ethanol also leads to the coproduction of several chemicals, from acetaldehyde to acetone and C_4 compounds, the relative amount of which is greatly affected by the spinel composition. The valorization of these compounds could help to render economically sustainable the industrial process.

Funding: This work was funded by SINCHEM Joint Doctorate Programme-Erasmus Mundus Action (framework agreement No. 2013-0037; specific grant agreement no. 2015-1600/001-001-EMJD).

Conflicts of Interest: The authors declare no conflicts of interest.

References

1. Jacobs, J.-P.; Maltha, A.; Reintjes, J.G.H.; Drimal, J.; Ponec, V.; Brongersma, H.H. The surface of catalytically active spinels. *J. Catal.* **1994**, *147*, 294–300. [CrossRef]
2. Zasada, F.; Piskorz, W.; Janas, J.; Gryboś, J.; Indyka, P.; Sojka, Z. Reactive Oxygen Species on the (100) Facet of Cobalt Spinel Nanocatalyst and their Relevance in $^{16}O_2/^{18}O_2$ Isotopic Exchange, *de*N$_2$O, and *de*CH$_4$ Processes. A Theoretical and Experimental Account. *ACS Catal.* **2015**, *5*, 6879–6892. [CrossRef]
3. Voinov, M.A.; Sosa Pagan, J.O.; Morrison, E.; Smirnova, T.U.; Smirnov, A.I. Surface-Mediated Production of Hydroxyl Radicals as a Mechanism of Iron Oxide Nanoparticle Biotoxicity. *J. Am. Chem. Soc.* **2011**, *133*, 35–41. [CrossRef] [PubMed]
4. Zhao, W.; Zhong, Q.; Ding, J.; Deng, Z.; Guo, L.; Song, F. Enhanced catalytic ozonation over reduced spinel CoMn$_2$O$_4$ for NOx removal: Active site and mechanism analysis. *RSC Adv.* **2016**, *6*, 115213–115221. [CrossRef]
5. Pardeshi, S.K.; Pawar, R.Y. Optimization of reaction conditions in selective oxidation of styrene over fine crystallite spinel-type CaFe$_2$O$_4$ complex oxide catalyst. *Mater. Res. Bull.* **2010**, *5*, 609–615. [CrossRef]
6. Tong, J.; Bo, L.; Li, Z.; Lei, Z.; Xia, C. Magnetic CoFe$_2$O$_4$ nanocrystal: A novel and efficient heterogeneous catalyst for aerobic oxidation of cyclohexane. *J. Mol. Catal. A Chem.* **2009**, *1–2*, 58–63. [CrossRef]
7. Kooti, M.; Afshari, M. Magnetic cobalt ferrite nanoparticles as an efficient catalyst for oxidation of alkenes. *Sci. Iran.* **2012**, *6*, 1991–1995. [CrossRef]
8. Yuan, C.; Wu, H.B.; Yi Xie, Y.; Lou, X.W. Mixed Transition-Metal Oxides: Design, Synthesis, and Energy-Related Applications. *Angew. Chem. (Int. Ed.)* **2014**, *53*, 1488–1504. [CrossRef] [PubMed]
9. Ati, A.A.; Othaman, Z.; Samavati, A.; Doust, F.Y. Structural and magnetic properties of Co–Al substituted Ni ferrites synthesized by coprecipitation method. *J. Mol. Struct.* **2014**, *1058*, 136–141. [CrossRef]
10. Ayyappan, S.; Philip, J.; Raj, B. A facile method to control the size and magnetic properties of CoFe$_2$O$_4$ nanoparticles. *Mater. Chem. Phys.* **2009**, *115*, 712–717. [CrossRef]
11. Chen, L.; Horiuchi, T.; Mori, T. Catalytic reduction of NO over a mechanical mixture of NiGa$_2$O$_4$ spinel with manganese oxide: Influence of catalyst preparation method. *Appl. Catal. A* **2001**, *209*, 97–105. [CrossRef]

12. Chia, C.H.; Zakaria, S.; Yusoff, M.; Goh, S.C.; Haw, C.Y.; Ahmadi, S.; Huang, N.M.; Lim, H.N. Size and crystallinity-dependent magnetic properties of CoFe$_2$O$_4$ nanocrystals. *Ceram. Int.* **2010**, *36*, 605–609. [CrossRef]

13. Chinnasamy, C.; Senoue, M.; Jeyadevan, B.; Perales-Perez, O.; Shinoda, K.; Tohji, K. Synthesis of size-controlled cobalt ferrite particles with high coercivity and squareness ratio. *J. Colloid Interface Sci.* **2003**, *263*, 80–83. [CrossRef]

14. Hirano, M.; Okumura, S.; Hasegawa, Y.; Inagaki, M. Direct precipitation of spinel type oxide ZnGa$_2$O$_4$ from aqueous solutions at low temperature below 90 °C. *Int. J. Inorg. Chem.* **2001**, *3*, 797–801. [CrossRef]

15. Hirano, M.; Okumura, S.; Hasegawa, Y.; Inagaki, M. Direct Precipitation of Spinel-Type Zn(Fe, Ga)$_2$O$_4$ Solid Solutions from Aqueous Solutions at 90 °C: Influence of Iron Valence of Starting Salt on Their Crystallite Growth. *J. Solid State Chem.* **2002**, *168*, 5–10. [CrossRef]

16. Liu, Z.L.; Wang, H.B.; Lu, Q.H.; Du, G.H.; Peng, L.; Du, Y.Q.; Zhang, S.M.; Yao, K.L. Synthesis and characterization of ultrafine well-dispersed magnetic nanoparticles. *J. Magn. Magn. Mater.* **2004**, *283*, 258–262. [CrossRef]

17. Paike, V.V.; Niphadkar, P.S.; Bokade, V.V.; Joshi, P.N. Synthesis of Spinel CoFe$_2$O$_4$ Via the Coprecipitation Method Using Tetraalkyl Ammonium Hydroxides as Precipitating Agents. *J. Am. Ceram. Soc.* **2007**, *90*, 3009–3012. [CrossRef]

18. Pereira, C.; Pereira, A.M.; Fernandes, C.; Rocha, M.; Mendes, R.; Fernández-García, M.P.; Guedes, A.; Tavares, P.B.; Grenèche, J.-M.; Araújo, J.P.; et al. Superparamagnetic MFe$_2$O$_4$ (M = Fe, Co, Mn) Nanoparticles: Tuning the Particle Size and Magnetic Properties through a Novel One-Step Coprecipitation Route. *Chem. Mater.* **2012**, *24*, 1496–1504. [CrossRef]

19. Prabhakar Vattikuti, S.V.; Byon, C.; Shim, J.; Reddy, C.V. Effect of temperature on structural, morphological and magnetic properties of Cd$_{0.7}$Co$_{0.3}$Fe$_2$O$_4$ nanoparticles. *J. Magn. Magn. Mater.* **2015**, *393*, 132–138. [CrossRef]

20. Qu, Y.; Yang, H.; Yang, N.; Fan, Y.; Zhu, H.; Zou, G. The effect of reaction temperature on the particle size, structure and magnetic properties of coprecipitated CoFe$_2$O$_4$ nanoparticles. *Mater. Lett.* **2006**, *60*, 3548–3552. [CrossRef]

21. Rafiq, M.A.; Khan, M.A.; Asghar, M.; Ilyas, S.Z.; Shakir, I.; Shahid, M.; Warsi, M.F. Influence of Co^{2+} on structural and electromagnetic properties of Mg–Zn nanocrystals synthesized via coprecipitation route. *Ceram. Int.* **2015**, *41*, 10501–10505. [CrossRef]

22. Singh, S.; Yadav, B.C.; Prakash, R.; Bajaj, B.; Lee, J.R. Synthesis of nanorods and mixed shaped copper ferrite and their applications as liquefied petroleum gas sensor. *Appl. Surf. Sci.* **2011**, *257*, 10763–10770. [CrossRef]

23. Soler, M.A.G.; Lima, E.C.D.; Silva, W.; Melo, T.F.O.; Pimenta, A.C.M.; Sinnecker, P.; Azevedo, R.B.; Garg, V.K.; Oliveira, A.C.; Novak, M.A.; et al. Aging Investigation of Cobalt Ferrite Nanoparticles in Low pH Magnetic Fluid. *Langmuir* **2007**, *23*, 9611–9617. [CrossRef] [PubMed]

24. Atif, M.; Hasanain, S.K.; Nadeem, M. Magnetization of sol–gel prepared zinc ferrite nanoparticles: Effects of inversion and particle size. *Solid State Commun.* **2006**, *138*, 416–421. [CrossRef]

25. Chae, K.P.; Lee, J.-G.; Su Kweon, H.; Bae Lee, Y. The crystallographic, magnetic properties of Al, Ti doped CoFe$_2$O$_4$ powders grown by sol–gel method. *J. Magn. Magn. Mater.* **2004**, *283*, 103–108. [CrossRef]

26. George, M.; Mary, J.A.; Nair, S.S.; Joy, P.A.; Anantharaman, M.R. Finite size effects on the structural and magnetic properties of sol–gel synthesized NiFe$_2$O$_4$ powders. *J. Magn. Magn. Mater.* **2006**, *302*, 190–195. [CrossRef]

27. Singhal, S.; Sharma, R.; Namgyal, T.; Jauhar, S.; Bhukal, S.; Kaur, J. Structural, electrical and magnetic properties of Co$_{0.5}$Zn$_{0.5}$Al$_x$Fe$_{2-x}$O$_4$ (x = 0, 0.2, 0.4, 0.6, 0.8 and 1.0) prepared via sol–gel route. *Ceram. Int.* **2012**, *38*, 2773–2778. [CrossRef]

28. Srivastava, M.; Ojha, A.K.; Chaubey, S.; Materny, A. Synthesis and optical characterization of nanocrystalline NiFe$_2$O$_4$ structures. *J. Alloys Compd.* **2009**, *481*, 515–519. [CrossRef]

29. Wen, J.; Ge, X.; Liu, X. Preparation of spinel-type Cd$_{1-x}$Mg$_x$Ga$_2$O$_4$ gas-sensing material by sol–gel method. *Sens. Actuators B Chem.* **2006**, *115*, 622–625. [CrossRef]

30. Xu, Y.; Wei, J.; Yao, J.; Fu, J.; Xue, D. Synthesis of CoFe$_2$O$_4$ nanotube arrays through an improved sol–gel template approach. *Mater. Lett.* **2008**, *62*, 1403–1405. [CrossRef]

31. Zhang, M.; Zi, Z.; Liu, Q.; Zhang, P.; Tang, X.; Yang, J.; Zhu, X.; Sun, Y.; Dai, J. Size Effects on Magnetic Properties of $Ni_{0.5}Zn_{0.5}Fe_2O_4$ Prepared by Sol–gel Method. *Adv. Mater. Sci. Eng.* **2013**, *2013*, 609819. [CrossRef]

32. Yan, K.; Wu, X.; An, X.; Xie, X. Facile synthesis and catalytic property of spinel ferrites by a template method. *J. Alloys Compd.* **2013**, *552*, 405–408. [CrossRef]

33. Barroso, M.N.; Gomez, M.F.; Gamboa, J.A.; Arrúa, L.A.; Abello, M.C. Preparation and characterization of CuZnAl catalysts by citrate gel process. *J. Phys. Chem. Solids* **2006**, *67*, 1583–1589. [CrossRef]

34. Hwang, B.J.; Santhanam, R.; Liu, D.G. Characterization of nanoparticles of $LiMn_2O_4$ synthesized by citric acid sol–gel method. *J. Power Sources* **2001**, *98*, 443–446. [CrossRef]

35. Nguyet, D.T.T.; Duong, N.P.; Satoh, T.; Anh, L.N.; Hien, T.D. Temperature-dependent magnetic properties of yttrium iron garnet nanoparticles prepared by citrate sol–gel. *J. Alloys Compd.* **2012**, *541*, 18–22. [CrossRef]

36. Hao, Y.; Lai, Q.; Liu, D.; Xu, Z.; Ji, X. Synthesis by citric acid sol–gel method and electrochemical properties of $Li_4Ti_5O_{12}$ anode material for lithium-ion battery. *Mater. Chem. Phys.* **2005**, *94*, 382–387. [CrossRef]

37. Arabi, H.; Ganjali, F. Structural and Magnetic Properties of Cobalt and Manganese Doped Ni-Ferrite Nanoparticles. *J. Supercond. Nov. Magn.* **2013**, *26*, 1031–1035. [CrossRef]

38. Concas, G.; Spano, G.; Cannas, C.; Musinu, A.; Peddis, D.; Piccaluga, G. Inversion degree and saturation magnetization of different nanocrystalline cobalt ferrites. *J. Magn. Magn. Mater.* **2009**, *321*, 1893–1897. [CrossRef]

39. Kanagesan, S.; Hashim, M.; Tamilselvan, S.; Alitheen, N.B.; Ismail, I. Sol–gel auto-combustion synthesis of cobalt ferrite and it's cytotoxicity properties. *Dig. J. Nanomater. Biostruct.* **2013**, *8*, 1601–1610.

40. Meng, Y.Y.; He, M.H.; Zeng, Q.; Jiao, D.L.; Shukla, S.; Ramanujan, R.V.; Liu, Z.W. Synthesis of barium ferrite ultrafine powders by a sol–gel combustion method using glycine gels. *J. Alloys Compd.* **2014**, *583*, 220–225. [CrossRef]

41. Shobana, M.K.; Sankar, S. Characterization of sol–gel-prepared nanoferrites. *J. Magn. Magn. Mater.* **2009**, *321*, 599–601. [CrossRef]

42. Shobana, M.K.; Sankar, S.; Rajendran, V. Structural and thermal studies of $Ni_{0.25}Mn_{0.75}Fe_2O_4$ composites by sol–gel combustion method. *J. Alloys Compd.* **2009**, *472*, 421–424. [CrossRef]

43. Sutka, A.; Mezinskis, G.; Pludons, A. Characterization of sol–gel auto-combustion derived spinel ferrite nano-materials. *Energetika* **2010**, *7*, 254–259. [CrossRef]

44. Version, I.; Gaffoor, A.; Ravinder, D. Characterization of Nano-Structured Nickel-Cobalt Ferrites Synthesized By Citrate-Gel Auto Combustion Method. *Int. J. Eng. Res. Appl.* **2014**, *4*, 73–79.

45. Xiao, S.H.; Jiang, W.F.; Li, L.Y.; Li, X.J. Low-temperature auto-combustion synthesis and magnetic properties of cobalt ferrite nanopowder. *Mater. Chem. Phys.* **2007**, *106*, 82–87. [CrossRef]

46. Agulhon, P.; Constant, S.; Lartigue, L.; Larionova, J.; Di Renzo, F.; Quignard, F. Controlled synthesis from alginate gels of cobalt–manganese mixed oxide nanocrystals with peculiar magnetic properties. *Catal. Today* **2012**, *189*, 49–54. [CrossRef]

47. Cai, Y.; Wang, H.; Yan, H.; Wang, B.; Yan, H.; Ahniyaz, A. Low-temperature hydrothermal synthesis of spinel-type lithium manganese oxide nanocrystallites. *Solid State Ion.* **2003**, *158*, 113–117. [CrossRef]

48. Gözüak, F.; Köseoğlu, Y.; Baykal, A.; Kavas, H. Synthesis and characterization of $Co_xZn_{1-x}Fe_2O_4$ magnetic nanoparticles via a PEG-assisted route. *J. Magn. Magn. Mater.* **2009**, *321*, 2170–2177. [CrossRef]

49. Guin, D.; Baruwati, B.; Manorama, S.V. A simple chemical synthesis of nanocrystalline AFe_2O_4 (A = Fe, Ni, Zn): An efficient catalyst for selective oxidation of styrene. *J. Mol. Catal. A Chem.* **2005**, *242*, 26–31. [CrossRef]

50. Hirano, M. Hydrothermal synthesis and characterization of $ZnGa_2O_4$ spinel fine particles. *J. Mater. Chem.* **2000**, *10*, 469–472. [CrossRef]

51. Liddle, B.J.; Collins, S.M.; Bartlett, B.M. A new one-pot hydrothermal synthesis and electrochemical characterization of $Li_{1+x}Mn_{2-y}O_4$ spinel structured compounds. *Energy Environ. Sci.* **2010**, *3*, 1339–1346. [CrossRef]

52. Shan, A.; Wu, X.; Lu, J.; Chen, C.; Wang, R. Phase formations and magnetic properties of single crystal nickel ferrite ($NiFe_2O_4$) with different morphologies. *CrystEngComm* **2015**, *17*, 1603–1608. [CrossRef]

53. Hirano, M.; Sakaida, N. Hydrothermal Synthesis and Low-Temperature Sintering of Zinc Gallate Spinel Fine Particles. *J. Am. Ceram. Soc.* **2002**, *85*, 1145–1150. [CrossRef]

54. Yu, S.; Fujino, T.; Yoshimura, M. Hydrothermal synthesis of $ZnFe_2O_4$ ultrafine particles with high magnetization. *J. Magn. Magn. Mater.* **2003**, *256*, 420–424. [CrossRef]

55. Zhang, X.D.; Wu, Z.S.; Zang, J.; Li, D.; Zhang, Z.D. Hydrothermal synthesis and characterization of nanocrystalline Zn-Mn spinel. *J. Phys. Chem. Solids* **2007**, *68*, 1583–1590. [CrossRef]

56. Blanco-Gutierrez, V.; Climent-Pascual, E.; Torralvo-Fernandez, M.J.; Saez-Puche, R.; Fernandez-Diaz, M.T. Neutron diffraction study and superparamagnetic behavior of $ZnFe_2O_4$ nanoparticles obtained with different conditions. *J. Solid State Chem.* **2011**, *184*, 1608–1613. [CrossRef]

57. Li, W.; Qiao, X.; Zheng, Q.; Zhang, T. One-step synthesis of MFe_2O_4 (M=Fe, Co) hollow spheres by template-free solvothermal method. *J. Alloys Compd.* **2011**, *509*, 6206–6211. [CrossRef]

58. Liu, J.; Zhang, Y.; Nan, Z. Facile synthesis of stoichiometric zinc ferrite nanocrystal clusters with superparamagnetism and high magnetization. *Mater. Res. Bull.* **2014**, *60*, 270–278. [CrossRef]

59. Yáñez-Vilar, S.; Sánchez-Andújar, M.; Gómez-Aguirre, C.; Mira, J.; Señarís-Rodríguez, M.A.; Castro-García, S. A simple solvothermal synthesis of MFe_2O_4 (M = Mn, Co and Ni) nanoparticles. *J. Solid State Chem.* **2009**, *182*, 2685–2690. [CrossRef]

60. Zawadzki, M. Pd and $ZnAl_2O_4$ nanoparticles prepared by microwave-solvothermal method as catalyst precursors. *J. Alloys Compd.* **2007**, *439*, 312–320. [CrossRef]

61. Choi, E.J.; Ahn, Y.; Song, K.C. Mössbauer study in zinc ferrite nanoparticles. *J. Magn. Magn. Mater.* **2006**, *301*, 171–174. [CrossRef]

62. Palmqvist, A.E.C. Synthesis of ordered mesoporous materials using surfactant liquid crystals or micellar solutions. *Curr. Opin. Colloid Interface Sci.* **2003**, *8*, 145–155. [CrossRef]

63. Ahn, Y.; Choi, E.J. Magnetization and Mossbauer Study of Nanosize $ZnFe_2O_4$ Particles Synthesized by Using a Microemulsion Method. *J. Korean Phys. Soc.* **2002**, *41*, 123–128.

64. Calero-DdelC, V.L.; Rinaldi, C. Synthesis and magnetic characterization of cobalt-substituted ferrite $(Co_xFe_{3-x}O_4)$ nanoparticles. *J. Magn. Magn. Mater.* **2007**, *314*, 60–67. [CrossRef]

65. Connor, C.J.O.; Seip, C.T.; Carpenter, E.E.; Li, S.; John, V.T. Synthesis and reactivity of nanophase in reverse micellar solutions ferrites. *Nanostruct. Mater.* **1999**, *12*, 65–70. [CrossRef]

66. Dar, M.A.; Shah, J.; Siddiqui, W.A.; Kotnala, R.K. Influence of synthesis approach on structural and magnetic properties of lithium ferrite nanoparticles. *J. Alloys Compd.* **2012**, *523*, 36–42. [CrossRef]

67. Lee, Y.; Lee, J.; Bae, C.J.; Park, J.-G.; Noh, H.-J.; Park, J.-H.; Hyeon, T. Large-Scale Synthesis of Uniform and Crystalline Magnetite Nanoparticles Using Reverse Micelles as Nanoreactors under Reflux Conditions. *Adv. Funct. Mater.* **2005**, *15*, 503–509. [CrossRef]

68. Liu, C.; Zou, B.; Rondinone, A.J.; Zhang, Z.J. Reverse Micelle Synthesis and Characterization of Superparamagnetic $MnFe_2O_4$ Spinel Ferrite Nanocrystallites. *J. Phys. Chem. B* **2000**, *104*, 1141–1145. [CrossRef]

69. Gu, M.; Yue, B.; Bao, R.; He, H. Template synthesis of magnetic one-dimensional nanostructured spinel MFe_2O_4 (M = Ni, Mg, Co). *Mater. Res. Bull.* **2009**, *44*, 1422–1427. [CrossRef]

70. Mitra, A.; Vázquez-Vázquez, C.; López-Quintela, M.A.; Paul, B.K.; Bhaumik, A. Soft-templating approach for the synthesis of high surface area and superparamagnetic mesoporous iron oxide materials. *Microporous Mesoporous Mater.* **2010**, *131*, 373–377. [CrossRef]

71. Thomas, A.; Premlal, B.; Eswaramoorthy, M. Synthesis of mesoporous Zn–Al spinel oxide nanorods with membrane like morphology. *Mater. Res. Bull.* **2006**, *41*, 1008–1014. [CrossRef]

72. Yuan, J.J.; Zhao, Q.; Xu, Y.S.; Liu, Z.G.; Du, X.B.; Wen, G.H. Synthesis and magnetic properties of spinel $CoFe_2O_4$ nanowire arrays. *J. Magn. Magn. Mater.* **2009**, *321*, 2795–2798. [CrossRef]

73. Azizi, A.; Sadrnezhaad, S.K. Effects of annealing on phase evolution, microstructure and magnetic properties of mechanically synthesized nickel-ferrite. *Ceram. Int.* **2010**, *36*, 2241–2245. [CrossRef]

74. Bid, S.; Pradhan, S.K. Preparation of zinc ferrite by high-energy ball-milling and microstructure characterization by Rietveld's analysis. *Mater. Chem. Phys.* **2003**, *82*, 27–37. [CrossRef]

75. Hajalilou, A.; Hashim, M.; Ebrahimi-Kahrizsangi, R.; Kamari, H.M.; Sarami, N. Synthesis and structural characterization of nano-sized nickel ferrite obtained by mechanochemical process. *Ceram. Int.* **2014**, *40*, 5881–5887. [CrossRef]

76. Hajalilou, A.; Hashim, M.; Kamari, H.M.; Masoudi, M.T. Effects of Milling Atmosphere and Increasing Sintering Temperature on the Magnetic Properties of Nanocrystalline $Ni_{0.36}Zn_{0.64}Fe_2O_4$. *J. Nanomater.* **2015**, 615739. [CrossRef]

77. Jiang, J.S.; Yang, X.L.; Gao, L.; Guo, J.K.; Jiang, J.Z. Synthesis and characterisation of nanocrystalline zinc ferrite. *J. Nanomater.* **1999**, *12*, 143–146. [CrossRef]

78. Manova, E.; Paneva, D.; Kunev, B.; Estournès, C.; Rivière, E.; Tenchev, K.; Léaustic, A.; Mitov, I. Mechanochemical synthesis and characterization of nanodimensional iron-cobalt spinel oxides. *J. Alloys Compd.* **2009**, *485*, 356–361. [CrossRef]

79. Manova, E.; Kunev, B.; Paneva, D.; Mitov, I.; Petrov, L.; Estourne, C.; D'Orlean, C.; Rehspringer, J.-L.; Kurmoo, M. Mechano-Synthesis, Characterization, and Magnetic Properties of Nanoparticles of Cobalt Ferrite, $CoFe_2O_4$. *Chem. Mater.* **2004**, *16*, 5689–5696. [CrossRef]

80. Šepelák, V.; Menzel, M.; Bergmann, I.; Wiebcke, M.; Krumeich, F.; Becker, K.D. Structural and magnetic properties of nanosize mechanosynthesized nickel ferrite. *J. Magn. Magn. Mater.* **2004**, *272–276*, 1616–1618. [CrossRef]

81. Vasoya, N.H.; Vanpariya, L.H.; Sakariya, P.N.; Timbadiya, M.D.; Pathak, T.K.; Lakhani, V.K.; Modi, K.B. Synthesis of nanostructured material by mechanical milling and study on structural property modifications in $Ni_{0.5}Zn_{0.5}Fe_2O_4$. *Ceram. Int.* **2010**, *36*, 947–954. [CrossRef]

82. Nawale, A.B.; Kanhe, N.S.; Patil, K.R.; Bhoraskar, S.V.; Mathe, V.L.; Das, A.K. Magnetic properties of thermal plasma synthesized nanocrystalline nickel ferrite ($NiFe_2O_4$). *J. Alloys Compd.* **2011**, *509*, 4404–4413. [CrossRef]

83. Velinov, N.; Manova, E.; Tsoncheva, T.; Estournès, C.; Paneva, D.; Tenchev, K.; Petkova, V.; Koleva, K.; Kunev, B.; Mitov, I. Spark plasma sintering synthesis of $Ni_{1-x}Zn_xFe_2O_4$ ferrites: Mössbauer and catalytic study. *Solid State Sci.* **2012**, *14*, 1092–1099. [CrossRef]

84. Nakatsuka, A.; Ikeda, Y.; Nakayama, N.; Mizota, T. Inversion parameter of the $CoGa_2O_4$ spinel determined from single-crystal X-ray data. *Acta Crystallogr. Sect. E* **2006**, *62*, i109–i111. [CrossRef]

85. Wang, W.H.; Ren, X. Flux growth of high-quality $CoFe_2O_4$ single crystals and their characterization. *J. Cryst. Growth* **2006**, *289*, 605–608. [CrossRef]

86. Yan, Z.; Takei, H.; Kawazoe, H. Electrical Conductivity in Transparent $ZnGa_2O_4$: Reduction and Surface-Layer Structure Transformation. *J. Am. Ceram. Soc.* **1998**, *86*, 180–186. [CrossRef]

87. Tasca, J.E.; Quincoces, C.E.; Lavat, A.; Alvarez, A.M.; González, M.G. Preparation and characterization of $CuFe_2O_4$ bulk catalysts. *Ceram. Int.* **2011**, *37*, 803–812. [CrossRef]

88. Kim, J.Y.; Kang, J.H.; Lee, D.C.; Jeon, D.Y. Preparation and characterization of $ZnGa_2O_4$ phosphor synthesized with an optimized combustion process. *J. Vac. Sci. Technol. B Microelectron. Nanom. Struct.* **2003**, *21*, 532. [CrossRef]

89. Zhang, X.; Jiang, W.; Song, D.; Sun, H.; Sun, Z.; Li, F. Salt-assisted combustion synthesis of highly dispersed superparamagnetic $CoFe_2O_4$ nanoparticles. *J. Alloys Compd.* **2009**, *475*, L34–L37. [CrossRef]

90. Zamudio, M.A.; Bensaid, S.; Fino, D.; Russo, N. Influence of the $MgCo_2O_4$ Preparation Method on N2O Catalytic Decomposition. *Ind. Eng. Chem. Res.* **2011**, *50*, 2622–2627. [CrossRef]

91. Ragupathi, C.; Vijaya, J.J.; Kennedy, L.J.; Bououdina, M. Nanostructured copper aluminate spinels: Synthesis, structural, optical, magnetic, and catalytic properties. *Mater. Sci. Semicond. Process.* **2014**, *24*, 146–156. [CrossRef]

92. Baykal, A.; Kasapoğlu, N.; Köseoğlu, Y.; Başaran, A.C.; Kavas, H.; Toprak, M.S. Microwave-induced combustion synthesis and characterization of $Ni_xCo_{1-x}Fe_2O_4$ nanocrystals (x = 0.0, 0.4, 0.6, 0.8, 1.0). *Cent. Eur. J. Chem.* **2008**, *6*, 125–130. [CrossRef]

93. Mahmoud, M.H.; Elshahawy, A.M.; Makhlouf, S.A.; Hamdeh, H.H. Synthesis of highly ordered 30 nm $NiFe_2O_4$ particles by the microwave-combustion method. *J. Magn. Magn. Mater.* **2014**, *369*, 55–61. [CrossRef]

94. Khollam, Y.B.; Dhage, S.R.; Potdar, H.S.; Deshpande, S.B. Microwave hydrothermal preparation of submicron-sized spherical magnetite (Fe_3O_4) powders. *Mater. Lett.* **2002**, *56*, 571–577. [CrossRef]

95. Kim, C.-K.; Lee, J.-H.; Katoh, S.; Murakami, R.; Yoshimura, M. Synthesis of Co-, Co-Zn and Ni-Zn ferrite powders by the microwave-hydrothermal method. *Mater. Res. Bull.* **2001**, *36*, 2241–2250. [CrossRef]

96. Lee, J.-H.; Kim, C.-K.; Katoh, S.; Murakami, R. Microwave-hydrothermal versus conventional hydrothermal preparation of Ni- and Zn-ferrite powders. *J. Alloys Compd.* **2001**, *325*, 276–280. [CrossRef]

97. Borges, F.M.M.; Melo, D.M.A.; Câmara, M.S.A.; Martinelli, A.E.; Soares, J.M.; de Araújo, J.H.; Cabral, F.A.O. Magnetic behavior of nanocrystalline $MnCo_2O_4$ spinels. *J. Magn. Magn. Mater.* **2006**, *302*, 273–277. [CrossRef]

98. De Souza Gonçalves, A.; Antonio Marques de Lima, S.; Rosaly Davolos, M.; Gutierrez Antônio, S.; de Oliveira Paiva-Santos, C. The effects of $ZnGa_2O_4$ formation on structural and optical properties of ZnO:Ga powders. *J. Solid State Chem.* **2006**, *179*, 1330–1334. [CrossRef]

99. Candeia, R.A.; Bernardi, M.I.B.; Longo, E.; Santos, I.M.G.; Souza, A.G. Synthesis and characterization of spinel pigment CaFe$_2$O$_4$ obtained by the polymeric precursor method. *Mater. Lett.* **2004**, *58*, 569–572. [CrossRef]

100. Pimentel, P.M.; Martinelli, A.E.; Melo, D.M.D.A.; Pedrosa, A.M.G.; Cunha, J.D.; Da Silva Júnior, C.N. Pechini synthesis and microstructure of nickel-doped copper chromites. *Mater. Res.* **2005**, *8*, 221–224. [CrossRef]

101. Galindo, R.; Mazario, E.; Gutiérrez, S.; Morales, M.P.; Herrasti, P. Electrochemical synthesis of NiFe$_2$O$_4$ nanoparticles: Characterization and their catalytic applications. *J. Alloys Compd.* **2012**, *536*, S241–S244. [CrossRef]

102. Ju, Y.-W.; Park, J.-H.; Jung, H.-R.; Cho, S.-J.; Lee, W.-J. Fabrication and characterization of cobalt ferrite (CoFe$_2$O$_4$) nanofibers by electrospinning. *Mater. Sci. Eng. B* **2008**, *147*, 7–12. [CrossRef]

103. Goodarz Naseri, M.; Saion, E.B.; Abbastabar Ahangar, H.; Shaari, A.H.; Hashim, M. Simple Synthesis and Characterization of Cobalt Ferrite Nanoparticles by a Thermal Treatment Method. *J. Nanomater.* **2012**, *2010*, 1–8. [CrossRef]

104. Naseri, M.G.; Saion, E.B.; Hashim, M.; Shaari, A.H.; Ahangar, H.A. Synthesis and characterization of zinc ferrite nanoparticles by a thermal treatment method. *Solid State Commun.* **2011**, *151*, 1031–1035. [CrossRef]

105. Chen, D.; Liu, H. One-step synthesis of nickel ferrite nanoparticles by ultrasonic wave-assisted ball milling technology. *Mater. Lett.* **2012**, *72*, 95–97. [CrossRef]

106. Jung, D.S.; Jang, H.C.; Lee, M.; Jung, K.Y.; Kang, Y.C. Synthesis and characterization of NiFe$_2$O$_4$ nanopowders via spray pyrolysis. *J. Ceram. Soc. Jpn.* **2009**, *117*, 1069–1073. [CrossRef]

107. Singhal, S.; Singh, J.; Barthwal, S.; Chandra, K. Preparation and characterization of nanosize nickel-substituted cobalt ferrites (Co$_{1-x}$Ni$_x$Fe$_2$O$_4$). *J. Solid State Chem.* **2005**, *178*, 3183–3189. [CrossRef]

108. Hanh, N.; Quy, O.K.; Thuy, N.P.; Tung, L.D.; Spinu, L. Synthesis of cobalt ferrite nanocrystallites by the forced hydrolysis method and investigation of their magnetic properties. *Phys. B Condens. Matter* **2003**, *327*, 382–384. [CrossRef]

109. Abdallah, H.M.I.; Moyo, T.; Ezekiel, I.P.; Osman, N.S.E. Structural and magnetic properties of Sr$_{0.5}$Co$_{0.5}$Fe$_2$O$_4$ nanoferrite. *J. Magn. Magn. Mater.* **2014**, *365*, 83–87. [CrossRef]

110. Jiang, J.; Yang, Y.-M.; Li, L.-C. Surfactant-assisted synthesis of nanostructured NiFe$_2$O$_4$ via a refluxing route. *Mater. Lett.* **2008**, *62*, 1973–1975. [CrossRef]

111. Acharyya, S.S.; Ghosh, S.; Adak, S.; Sasakib, T.; Bal, R. Facile synthesis of CuCr$_2$O$_4$ spinel nanoparticles: A recyclable heterogeneous catalyst for the one pot hydroxylation of benzene. *Catal. Sci. Technol.* **2014**, *4*, 4232–4241. [CrossRef]

112. Sahaa, S.; Hamid, S.B.A. Nanosized spinel Cu–Mn mixed oxide catalyst prepared via solvent evaporation for liquid phase oxidation of vanillyl alcohol using air and H$_2$O$_2$. *RSC Adv.* **2016**, *6*, 96314–96326. [CrossRef]

113. Ragupathia, C.; Vijayaa, J.J.; Narayanana, S.; Jesudossa, S.K.; Kennedy, L.J. Highly selective oxidation of benzyl alcohol to benzaldehyde with hydrogen peroxide by cobalt aluminate catalysis: A comparison of conventional and microwave methods. *Ceram. Int.* **2015**, *41*, 2069–2080. [CrossRef]

114. Menini, L.; Pereira, M.C.; Parreira, L.A.; Fabris, J.D.; Gusevskaya, E.V. Cobalt- and manganese-substituted ferrites as efficient single-site heterogeneous catalysts for aerobic oxidation of monoterpenic alkenes under solvent-free conditions. *J. Catal.* **2008**, *254*, 355–364. [CrossRef]

115. Gawade, A.B.; Nakhate, A.V.; Yadav, G.D. Selective synthesis of 2, 5-furandicarboxylic acid by oxidation of 5-hydroxymethylfurfural over MnFe$_2$O$_4$ catalyst. *Catal. Today* **2018**, *309*, 119–125. [CrossRef]

116. Jain, A.; Jonnalagadda, S.C.; Ramanujachary, K.V.; Mugweru, A. Selective oxidation of 5-hydroxymethyl-2-furfural to furan-2,5-dicarboxylic acid over spinel mixed metal oxide catalyst. *Catal. Commun.* **2015**, *58*, 179–182. [CrossRef]

117. Mishra, D.K.; Lee, H.J.; Kim, J.; Lee, H.S.; Cho, J.K.; Suh, Y.-W.; Yid, Y.; Kim, Y.J. MnCo$_2$O$_4$ spinel supported ruthenium catalyst for air-oxidation of HMF to FDCA under aqueous phase and base-free conditions. *Green Chem.* **2017**, *19*, 1619–1623. [CrossRef]

118. Acharyya, S.S.; Ghosh, S.; Bal, R. Catalytic Oxidation of Aniline to Azoxybenzene Over CuCr$_2$O$_4$ Spinel Nanoparticle Catalyst. *ACS Sustain. Chem. Eng.* **2014**, *2*, 584–589. [CrossRef]

119. Acharyya, S.S.; Ghosh, S.; Tiwari, V.; Sarkar, B.; Singha, R.K.; Pendem, C.; Sasakib, T.; Bal, R. Preparation of the CuCr$_2$O$_4$ spinel nanoparticles catalyst for selective oxidation of toluene to benzaldehyde. *Green Chem.* **2014**, *16*, 2500–2508. [CrossRef]

120. Song, W.; Liua, P.; Hensen, E.J.M. A mechanism of gas-phase alcohol oxidation at the interface of Au nanoparticles and a $MgCuCr_2O_4$ spinel support. *Catal. Sci. Technol.* **2014**, *4*, 2997–3003. [CrossRef]

121. Mate, V.R.; Shirai, M.; Rode, C.V. Heterogeneous Co_3O_4 catalyst for selective oxidation of aqueous veratryl alcohol using molecular oxygen. *Catal. Comm.* **2013**, *33*, 66–69. [CrossRef]

122. Klimkiewicz, R.; Wolska, J.; Przepiera, A.; Przepiera, K.; Jabłoński, M.; Lenart, S. The zinc ferrite obtained by oxidative precipitation method as a catalyst in n-butanol conversion. *Mater. Res. Bull.* **2009**, *44*, 15–20. [CrossRef]

123. Behar, S.; Gómez-Mendoza, N.-A.; Gómez-García, M.-A.; Swierczynski, D.; Quignard, F.; Tanchoux, N. Study and modelling of kinetics of the oxidation of VOC catalyzed by nanosized Cu–Mn spinels prepared via an alginate route. *Catal. Today* **2012**, *189*, 35–41. [CrossRef]

124. Ghorpade, S.P.; Darshane, V.S.; Dixit, S.G. Liquid-phase Friedel-Crafts alkylation using $CuCr_{2-x}Fe_xO_4$ spinel catalysts. *Appl. Catal. A* **1998**, *166*, 135–142. [CrossRef]

125. Rostami, A.; Atashkar, B.; Gholami, H. Novel magnetic nanoparticles Fe_3O_4-immobilized domino Knoevenagel condensation, Michael addition, and cyclization catalyst. *Catal. Commun.* **2013**, *37*, 69–74. [CrossRef]

126. Kantam, M.L.; Yadav, J.; Laha, S.; Srinivas, P.; Sreedhar, B.; Figueras, F. Asymmetric Hydrosilylation of Ketones Catalyzed by Magnetically Recoverable and Reusable Copper Ferrite Nanoparticles. *J. Org. Chem.* **2009**, *74*, 4608–4611. [CrossRef] [PubMed]

127. Ibraheem, I.I.; Tarek, O.A.; Salamaa, M.; Bahgatb, A.A.; Mohamed, M.M. Synthesis of magnetically recyclable spinel ferrite (MFe_2O_4, M = Zn, Co, Mn) nanocrystals engineered by sol gel-hydrothermal technology: High catalytic performances for nitroarenes reduction. *Appl. Catal. B* **2016**, *181*, 389–402. [CrossRef]

128. Sreekumar, K.; Sugunan, S. Ferrospinels based on Co and Ni prepared via a low temperature route as efficient catalysts for the selective synthesis of o-cresol and 2,6-xylenol from phenol and methanol. *J. Mol. Catal. A Chem.* **2002**, *185*, 259–268. [CrossRef]

129. Calvillo, L.; Carraro, F.; Vozniuk, O.; Celorrio, V.; Nodari, L.; Russell, A.E.; Debellis, D.; Fermin, D.; Cavani, F.; Agnoli, S.; et al. Insights into the durability of Co–Fe spinel oxygen evolution electrocatalysts via operando studies of the catalyst structure. *J. Mater. Chem. A* **2018**, *6*, 7034–7041. [CrossRef]

130. Carraro, F.; Vozniuk, O.; Calvillo, L.; Nodari, L.; La Fontaine, C.; Cavani, F.; Agnoli, S. In operando XAS investigation of reduction and oxidation processes in cobalt and iron mixed spinels during the chemical loop reforming of ethanol. *J. Mater. Chem. A* **2017**, *5*, 20808–20817. [CrossRef]

131. Vozniuk, O.; Bazzo, C.; Albonetti, C.; Tanchoux, N.; Bosselet, F.; Millet, J-M.M.; Di Renzo, F.; Cavani, F. Structural Changes of Binary/Ternary Spinel Oxides During Ethanol Anaerobic Decomposition. *ChemCatChem* **2017**, *9*, 2219–2230. [CrossRef]

132. Trevisanut, C.; Vozniuk, O.; Mari, M.; Arenas Urrea, S.Y.; Lorentz, C.; Millet, J.-M.M.; Cavani, F. The Chemical-Loop Reforming of Alcohols on Spinel-Type Mixed Oxides: Comparing Ni, Co, and Fe Ferrite vs Magnetite Performances. *Top Catal.* **2016**, *59*, 1600–1613. [CrossRef]

133. Vozniuk, O.; Agnoli, S.; Artiglia, L.; Vassoi, A.; Tanchoux, N.; Di Renzo, F.; Granozzi, G.; Cavani, F. Towards an improved process for hydrogen production: The chemical-loop reforming of ethanol. *Green Chem.* **2016**, *18*, 1038–1050. [CrossRef]

134. Trevisanut, C.; Mari, M.; Millet, J.-M.M.; Cavani, F. Chemical-loop reforming of ethanol over metal ferrites: An analysis of structural features affecting Reactivity. *Int. J. Hydrogen Energy* **2015**, *40*, 5264–5271. [CrossRef]

135. Trevisanut, C.; Bosselet, F.; Cavani, F.; Millet, J.M.M. A study of surface and structural changes of magnetite cycling material during chemical looping for hydrogen production from bioethanol. *Catal. Sci. Technol.* **2015**, *5*, 1280–1289. [CrossRef]

136. Cocchi, S.; Mari, M.; Cavani, F.; Millet, J.-M.M. Chemical and physical behavior of $CoFe_2O_4$ in steam-iron process with methanol. *Appl. Catal. B* **2014**, *152–153*, 250–261. [CrossRef]

137. Crocellà, V.; Cavani, F.; Cerrato, G.; Cocchi, S.; Comito, M.; Magnacca, G.; Morterra, C. On the Role of Morphology of $CoFeO_4$ Spinel in Methanol Anaerobic Oxidation. *J. Phys. Chem. C* **2012**, *116*, 14998–15009. [CrossRef]

138. Velasquez Ochoa, J.; Trevisanut, C.; Millet, J.M.M.; Busca, G.; Cavani, F. In Situ DRIFTS-MS Study of the Anaerobic Oxidation of Ethanol over Spinel Mixed Oxides. *J. Phys. Chem. C* **2013**, *117*, 23908–23918. [CrossRef]

139. Protasova, L.; Snijkers, F. Recent developments in oxygen carrier materials for hydrogen production via chemical looping processes. *Fuel* **2016**, *181*, 75–93. [CrossRef]

140. Cha, K.-S.; Kim, H.-S.; Yoo, B.-K.; Lee, Y.-S.; Kang, K.-S.; Park, C.-S.; Kim, Y.-H. Reaction characteristics of two-step methane reforming over a Cu-ferrite/Ce–ZrO$_2$ medium. *Int. J. Hydrogen Energy* **2009**, *34*, 1801–1808. [CrossRef]

141. Kang, K.-S.; Kim, C.-H.; Cho, W.-C.; Bae, K.-K.; Woo, S.-W.; Park, C.-S. Reduction characteristics of CuFe$_2$O$_4$ and Fe$_3$O$_4$ by methane; CuFe$_2$O$_4$ as an oxidant for two-step thermochemical methane reforming. *Int. J. Hydrogen Energy* **2008**, *33*, 4560–4568. [CrossRef]

142. Yamaguchi, D.; Tang, L.; Wong, L.; Burke, N.; Trimm, D.; Nguyen, K.; Chiang, K. Hydrogen production through methane-steam cyclic redox processes with iron-based metal oxides. *Int. J. Hydrogen Energy* **2011**, *36*, 6646–6656. [CrossRef]

143. Takenaka, S.; Hanaizumi, N.; Son, V.; Otsuka, K. Production of pure hydrogen from methane mediated by the redox of Ni- and Cr-added iron oxides. *J. Catal.* **2004**, *228*, 405–416. [CrossRef]

144. Kang, K.-S.; Kim, C.-H.; Bae, K.-K.; Cho, W.-C.; Kim, W.-J.; Kim, Y.-H.; Kim, S.-H.; Park, C.-S. Redox cycling of CuFe$_2$O$_4$ supported on ZrO$_2$ and CeO$_2$ for two-step methane reforming/water splitting. *Int. J. Hydrogen Energy* **2010**, *35*, 568–576. [CrossRef]

145. Otsuka, K.; Kaburagi, T.; Yamada, C.; Takenaka, S. Chemical storage of hydrogen by modified iron oxides. *J. Power Sources* **2003**, *122*, 111–121. [CrossRef]

146. Lorente, E.; Peña, J.A.; Herguido, J. Cycle behaviour of iron ores in the steam-iron process. *Int. J. Hydrogen Energy* **2011**, *36*, 7043–7050. [CrossRef]

147. Takenaka, S.; Kaburagi, T.; Yamada, C.; Nomura, K.; Otsuka, K. Storage and supply of hydrogen by means of the redox of the iron oxides modified with Mo and Rh species. *J. Catal.* **2004**, *228*, 66–74. [CrossRef]

148. Takenaka, S.; Nomura, K.; Hanaizumi, N.; Otsuka, K. Storage and formation of pure hydrogen mediated by the redox of modified iron oxides. *Appl. Catal. A Gen.* **2005**, *282*, 333–341. [CrossRef]

149. Ryu, J.C.; Lee, D.H.; Kang, K.S.; Park, C.S.; Kim, J.W.; Kim, Y.H. Effect of additives on redox behavior of iron oxide for chemical hydrogen storage. *J. Ind. Eng. Chem.* **2008**, *14*, 252–260. [CrossRef]

150. Otsuka, K.; Yamada, C.; Kaburagi, T.; Takenaka, S. Hydrogen storage and production by redox of iron oxide for polymer electrolyte fuel cell vehicles. *Int. J. Hydrogen Energy* **2003**, *28*, 335–342. [CrossRef]

151. Aston, V.J.; Evanko, B.W.; Weimer, A.W. Investigation of novel mixed metal ferrites for pure H$_2$ and CO$_2$ production using chemical looping. *Int. J. Hydrogen Energy* **2013**, *38*, 9085–9096. [CrossRef]

152. Bhavsar, S.; Tackett, B.; Veser, G. Evaluation of iron- and manganese-based mono- and mixed-metallic oxygen carriers for chemical looping combustion. *Fuel* **2014**, *136*, 268–279. [CrossRef]

153. Chiron, F.-X.; Patience, G.S. Kinetics of mixed copper-iron based oxygen carriers for hydrogen production by chemical looping water splitting. *Int. J. Hydrogen Energy* **2012**, *37*, 10526–10538. [CrossRef]

154. Chiron, F.-X.; Patience, G.S.; Rifflart, S. Hydrogen production through chemical looping using NiO/NiAl$_2$O$_4$ as oxygen carrier. *Chem. Eng. Sci.* **2011**, *66*, 6324–6330. [CrossRef]

155. Cormos, C.-C. Evaluation of iron based chemical looping for hydrogen and electricity co-production by gasification process with carbon capture and storage. *Int. J. Hydrogen Energy* **2010**, *35*, 2278–2289. [CrossRef]

156. Go, K.; Son, S.; Kim, S.; Kang, K.; Park, C. Hydrogen production from two-step steam methane reforming in a fluidized bed reactor. *Int. J. Hydrogen Energy* **2009**, *34*, 1301–1309. [CrossRef]

157. Källén, M.; Hallberg, P.; Rydén, M.; Mattisson, T.; Lyngfelt, A. Combined oxides of iron, manganese and silica as oxygen carriers for chemical-looping combustion. *Fuel Process. Technol.* **2014**, *124*, 87–96. [CrossRef]

158. Lorentzou, S.; Zygogianni, A.; Tousimi, K.; Agrafiotis, C.; Konstandopoulos, A.G. Advanced synthesis of nanostructured materials for environmental applications. *J. Alloys Compd.* **2009**, *483*, 302–305. [CrossRef]

159. Luo, M.; Wang, S.; Wang, L.; Lv, M. Reduction kinetics of iron-based oxygen carriers using methane for chemical-looping combustion. *J. Power Sources* **2014**, *270*, 434–440. [CrossRef]

160. Rosmaninho, M.G.; Moura, F.C.C.; Souza, L.R.; Nogueira, R.K.; Gomes, G.M.; Nascimento, J.S.; Pereira, M.C.; Fabris, J.D.; Ardisson, J.D.; Nazzarro, M.S.; et al. Investigation of iron oxide reduction by ethanol as a potential route to produce hydrogen. *Appl. Catal. B Environ.* **2012**, *115–116*, 45–52. [CrossRef]

catalysts

MDPI

Review

Catalytic Transfer Hydrogenolysis as an Effective Tool for the Reductive Upgrading of Cellulose, Hemicellulose, Lignin, and Their Derived Molecules

Claudia Espro [1], Bianca Gumina [1], Tomasz Szumelda [2], Emilia Paone [3] and Francesco Mauriello [3,*]

1   Dipartimento di Ingegneria, Università di Messina, Contrada di Dio–Vill. S. Agata, I-98166 Messina, Italy; espro@unime.it (C.E.); bianca.gumina@unime.it (B.G.)
2   Jerzy Haber Institute of Catalysis and Surface Chemistry, Polish Academy of Sciences, Niezapominajek 8, 30-239 Krakow, Poland; ncszumel@cyf-kr.edu.pl
3   Dipartimento DICEAM, Università Mediterranea di Reggio Calabria, Loc. Feo di Vito, I-89122 Reggio Calabria, Italy; emilia.paone@unirc.it
*   Correspondence: francesco.mauriello@unirc.it; Tel.: +39-0965-169-2278

Received: 9 July 2018; Accepted: 28 July 2018; Published: 31 July 2018

Abstract: Lignocellulosic biomasses have a tremendous potential to cover the future demand of bio-based chemicals and materials, breaking down our historical dependence on petroleum resources. The development of green chemical technologies, together with the appropriate eco-politics, can make a decisive contribution to a cheap and effective conversion of lignocellulosic feedstocks into sustainable and renewable chemical building blocks. In this regard, the use of an indirect H-source for reducing the oxygen content in lignocellulosic biomasses and in their derived platform molecules is receiving increasing attention. In this contribution we highlight recent advances in the transfer hydrogenolysis of cellulose, hemicellulose, lignin, and of their derived model molecules promoted by heterogeneous catalysts for the sustainable production of biofuels and biochemicals.

Keywords: lignocellulosic biomasses; H-donor molecules; hydrogenolysis; catalytic transfer hydrogenolysis reactions; heterogeneous catalysis; cellulose; hemicellulose; lignin; glycerol; polyols; furfural; levulinic acid; aromatic ethers

1. Introduction

The hegemony of fossil resources is declining by now. In the last few decades, industrial chemistry has accepted the challenge for the sustainable production of chemicals and energy by using renewable biomasses as starting supplies [1]. Moreover, the changes in consumer attitudes towards "green" products, as well as government initiatives for sustainable development programs and regulations, are surely the key driving factors for the development of the bio-based chemical industries and refineries [2–6].

While many criticisms have been raised towards the first generation of bio-energies and biofuels since they are in direct competition with human and animal food, reducing the land availability [7], we have recently achieved significant progress in the production of chemical building blocks and intermediates from lignocellulosic wastes and residues [8–20]. This is because their use in the chemical industry presents several advantages including: (i) the production of less toxic by-products and lower environmental risks, (ii) the reduction of CO_2 emissions, (iii) a minor dependence on fossil resources and/or foreign commodities, and (iv) the use of indigenous raw materials that can add value in many agriculture products or processes.

Three different approaches have been used, so far, by the bio-based refineries: (i) the production of the identical petrochemical building block starting from the lignocellulosic platform molecules, (ii) the use of platform molecules to produce the first petrochemical intermediates, and (iii) to synthesize new products alternative to the petrochemical ones starting from the platform molecules and/or their intermediates. It is expected that the market relative to all bio-based chemicals and materials will increase at an annual growth rate of 16.53% in the forecast period 2018–2026 [21] with a total product value of about 103 billion euros by 2050 [22].

Natural abundance, renewability, and recyclability of non-edible lignocellulose-based wastes and residues make them one of the most eco-attractive and alternative candidates to replace unrenewable petroleum-based resources in modern industrial chemistry.

The chemical structure of lignocellulosic biomasses allows the production of a wide spectra of platform chemicals (Figure 1) [23].

Cellulose and hemicellulose allow the production of C5-C6 sugars that can be easily converted into aliphatic acids, ethers, esters, polyols, and alcohols [24–29], while lignin is a source of aromatic compounds [30–33].

Cellulose is a linear glucosidic-based polymer with a degree of polymerization of around 100,000 units containing 49 wt% of oxygen [23,34]. Hemicellulose has a heterogeneous architecture based on pentoses, hexoses, and sugar acids with a degree of polymerization ranging from 100 to 200 units in which anhydrous sugars alternate with five- and six-carbon atoms and an overall oxygen content of 54 wt% is present [23,34]. Lignin is a branched polymer mainly consisting of phenols whose oxygen content is between 12 and 29 wt% [23,34].

In order to lower the oxygen content in cellulose, hemicellulose, and lignin—as well as in their relative derived molecules—three different chemical ways can be followed: (i) the removal of small molecules of oxidized carbon (CO_2, formaldehyde, and formic acid), (ii) the elimination of water through dehydration processes, and (iii) the direct "lysis" of the C–O bond by molecular H_2 with the concurrent formation of a molecule of water.

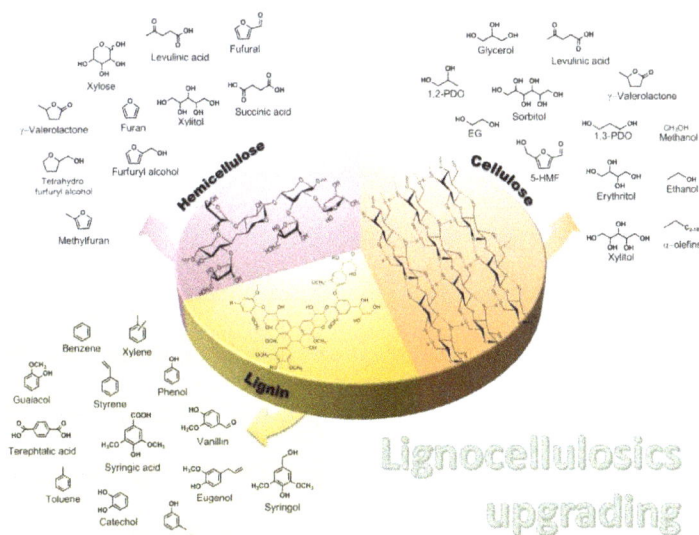

Figure 1. Chemical structure of cellulose, hemicellulose, lignin, and their relative derived molecules.

Hydrogenolysis is a very well-known reaction in which carbon–carbon or carbon–heteroatom bonds are cleaved, generally in the presence of a homogeneous or heterogeneous catalyst, that has been efficiently adopted for the reduction of the oxygen content in lignocellulosic components and in their derived platform chemicals [35–37].

The reductive valorization of lignocellulosic biomasses and their relative derived molecules are generally conducted in the presence of a solvent in order to limit their thermal decomposition. As a consequence, due to the well-known poor solubility of H_2 in most solvents, hydrogenolysis processes require the direct use of high-pressure molecular hydrogen with all concurrent problems that this entails, including purchase, transport, costly infrastructures, and safety hazards.

Simple organic molecules represent a valid green alternative to the direct use of molecular H_2 in reductive processes [38–40]. The ability of alcohols as potential sources of hydrogen in catalytic transfer hydrogenolysis (CTH) reactions can be correlated with their reduction potentials (defined as the difference of the standard molar free energy of formation between the alcohol and the corresponding carbonyl compound).

In CTH reactions, the use of H-donor molecules as solvent reduces the safety problems of handling high-pressure and explosive hydrogen gas, reducing, at the same time, costs and complexity of bio-based industrial chemical plants. Moreover, at present, many H-donor molecules can be easily produced from renewable feedstocks [41]. Nonetheless of importance, H-donor molecules, due to their lower hydrogenation ability with respect to molecular H_2, generally allow a higher production of aromatic compounds.

Formic acid was also efficiently used as a potential liquid storage medium capable of releasing H_2 under mild conditions via catalytic decomposition [42,43]. Formic acid can be formed by renewable technologies, namely from the lignocellulosic biomass or from electrochemical reduction of CO_2, which makes it an environmentally friendly source for both high-purity hydrogen production and hydrogen donor for CTH reactions [44].

In this review, we examine the recent progress in the CTH of cellulose, hemicellulose, lignin, and their relative derived molecules promoted by heterogeneous catalysts, focusing our attention in the C–O and C–C bond breaking. The overall aim is to offer a complete overview of the huge potential offered by the catalytic transfer hydrogenolysis reaction in the upgrading lignocellulosic resources for the sustainable production of biofuels and biochemicals.

2. Catalytic Transfer Hydrogenolysis Applied to Cellulose and to Cellulose Derivable Molecules

2.1. Glycerol and Other Polyols

Among the family of cellulose derived molecules, C6-C3 polyols are of particular interest, being used as a starting resource for several building block chemicals.

Glycerol is the main by-product in biodiesel manufacture and, at the same time, it is easily derivable from cellulose thus being a promising renewable molecule to obtain, among others, 1,2-propanediol (1,2-PDO) which is an important polymer precursor. Therefore, the conversion of glycerol into 1,2-PDO, through the catalytic transfer hydrogenolysis, becomes an interesting tool. The main results present in literature concerning several catalytic substrates, applied to the CTH of glycerol, are summarized in Table 1.

The catalytic transfer hydrogenolysis (CTH) of glycerol, was performed for the first time by Pietropaolo and co-workers, using 2-propanol (2-PO) as a hydrogen donor and solvent [45]. The investigation started by using the unreduced bimetallic catalyst PdO/Fe_2O_3 and reaches complete conversion of glycerol and a high selectivity of 94% towards 1,2-propanediol (1,2-PDO), after 24 h of reaction at 180 °C. Ethanol was also used showing slightly lower selectivity to 1,2-PDO (90%) in the same operating conditions.

The transfer hydrogenolysis of glycerol into 1,2-propanediol, catalyzed by bimetallic Pd/Fe_3O_4 and Pd/Co_3O_4 catalysts, enlightens a mechanism in which the glycerol (i) adsorbs over the bimetallic sites giving dehydration through the breaking a C–OH of the primary alcoholic group, and (ii) thanks to the hydrogen supplied from the 2-propanol dehydrogenation, the intermediate acetol can be hydrogenated into 1,2-propanediol (Scheme 1) [46]. Furthermore, a close correlation between the ability of catalysts towards the dehydrogenation of 2-propanol and the ability to perform CTH reactions was found [46].

Table 1. Catalytic transfer hydrogenolysis of glycerol to 1,2-propanediol.

Entry	Catalyst	H-Donor [1]	Cat/Gly [2]	Temp (°C)	Time (h)	Conv. (%)	1,2-PDO Select. (%)	Ref.
1	PdO/Fe_2O_3	EtOH	0.237	180	24	100	90	[45]
2	PdO/Fe_2O_3	2-PO	0.237	180	24	100	94	[45]
3	PdO/Fe_2O_3	2-PO	0.237	180	8	96	87	[45]
4	Pd/Fe_3O_4	2-PO	0.237	180	8	100	84	[45]
5	Pd/Fe_3O_4	2-PO	0.207	180	24	100	56	[46]
6	Pd/Co_3O_4	2-PO	0.207	180	24	100	64	[46]
7	$Ni-Cu/Al_2O_3$	-	0.166	220	24	70.5	66.9	[47]
8	$Ni-Cu/Al_2O_3$	2-PO	0.166	220	24	60.4	64.6	[47]
9	$Ni-Cu/Al_2O_3$	2-PO	0.166	220	10	28.2	77.4	[48]
10	$Ni-Cu/Al_2O_3$	MeOH	0.120	220	10	26.2	51.2	[48]
11	$Ni-Cu/Al_2O_3$	FA	0.120	220	10	33.5	85.9	[48]
12	$Ni-Cu/Al_2O_3$	FA	0.498	220	24	90	82	[49]
13	70Cu30Al	2-PO	-	220	5	69	90	[50]
14	$20Cu/ZrO_2$	FA	-	220	18	97	95	[51]

[1] Abbreviations: EtOH: ethanol; 2-PO: 2-propanol; MeOH: methanol; FA: formic acid; [2] Ratio of Cat/Gly (g/g).

Scheme 1. Mechanism of CTH of glycerol over Pd/Fe_3O_4 and Pd/Co_3O_4 catalysts. Adapted from Ref. [46]. Copyright Year 2015, Elsevier.

Gandarias et al. have deeply studied the reactivity of bimetallic catalysts $Ni-Cu/Al_2O_3$, prepared through the sol-gel technique, giving the best performances when pretreated at 450 °C [47–49]. At the beginning, the hydrogenolysis of glycerol was studied either in presence of molecular hydrogen or in conditions in which the hydrogen is produced in situ, through the aqueous phase reforming (APR) or through the catalytic transfer hydrogenolysis (CTH) by means of 2-propanol [47]. Results suggest that 2-propanol is a more effective hydrogen source than the aqueous-phase reforming, for the glycerol hydrogenolysis process. However, CTH of 2-propanol data are comparable to those obtained in the presence of molecular hydrogen. Two different mechanisms are involved when the hydrogenolysis of glycerol is performed in the presence of molecular hydrogen or in that of 2-PO (Scheme 2). Moreover, the deactivation of the catalyst occurs more rapidly in the presence of 2-PO, because adjacent sites are required for the donor and the acceptor processes relative to the transfer reaction, during the hydrogenation [47].

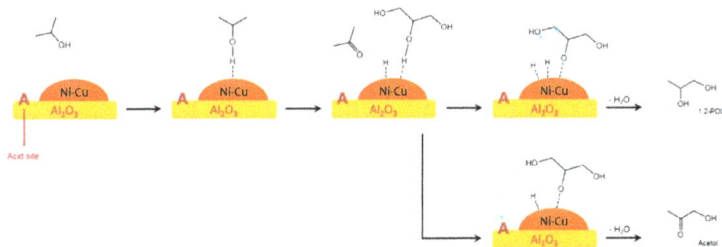

Scheme 2. Mechanism of CTH of glycerol over $Ni-Cu/Al_2O_3$ catalyst. Adapted from Ref. [47]. Copyright Year 2011, Elsevier.

On the other hand, by changing the hydrogen donor molecule, it is possible to improve the performance of the CTH of glycerol on using the bimetallic Ni-Cu/Al$_2$O$_3$ catalyst [49]. Gandarias et al. performed a study on the effect of the hydrogen donor molecule, by comparing methanol (MeOH), formic acid (FA) and 2-propanol [48]. Formic acid appears to be the most effective hydrogen donor molecule. The scale of efficiency, per donor molecule, follows this order: formic acid > 2-propanol > methanol. With the aim to further improve the reactivity of the Ni-Cu/Al$_2$O$_3$ catalyst, it is necessary to add formic acid and molecular hydrogen in order to obtain a conversion of glycerol of 43.9% and a selectivity to 1,2-PDO of almost 90% [48]. The kinetic study enables to understand that the hydroxyl groups of glycerol and the target product 1,2-PDO adsorb over the same acidic sites of the Al$_2$O$_3$ support. Therefore, in order to overcome this problem, it is necessary to increase the amount of catalyst and high conversion (90%) and selectivity (82%), within 24 h of reaction, were obtained [49].

Furthermore, the higher prevalence of the C–O bond cleavage is produced by a Cu-Ni alloy, whereas active Ni ensembles are responsible for both the C–C and the C–O bond cleavage, and the Cu ensembles favor mainly the C–O breaking but not the C–C one. Therefore, in the case of the reduced Ni-Cu/Al$_2$O$_3$ catalyst, the C–C bond cleavage is limited, while the activity to the C–O bond cleavage is significantly promoted [49].

Rasika et al. performed a study on the dehydration and the hydrogenolysis of glycerol by using both water and 2-propanol and screening the reactivity of Cr-based and Cu-Al catalysts [50]. Considering the hydrogenolysis of glycerol, the co-precipitated Cu-Al catalysts have shown a better performance, particularly in conditions of the CTH promoted by 2-propanol. Indeed, the 70Cu30Al catalyst reaches 69% of conversion and 90% of selectivity in 1,2-PDO within 5h of reaction. The evidence of different performances in the presence of 2-propanol instead of water indicates that the two processes are ruled by two different reaction pathways [50].

Yuan et al. investigated a series of Cu/ZrO$_2$-based catalysts, particularly the 20%Cu/ZrO$_2$, synthesized by co-precipitation, and found they can be used to convert glycerol to 1,2-PDO in high yields with formic acid as a hydrogen source. They found that the production of 1,2-PDO can be optimized using a FA/glycerol molar ratio (1:1) and a temperature of 200°C. In this case, a yield of 1,2-PDO (94%) after 18 h is obtained [51]. Furthermore, this catalyst is also pretty stable, and reusable at least five times without losing any appreciable reactivity and selectivity [51].

In the literature, some works that report the conversion of glycerol in traditional hydrogenolysis conditions (with addition of molecular hydrogen), but operating in a solvent that can donate hydrogen, like ethanol and 2-propanol, are present (Table 2).

A bimetallic Pd-Cu/solid-base catalyst was prepared via thermal decomposition of Pd$_x$Cu$_{0.4}$Mg$_{5.6-x}$Al$_2$(OH)$_{16}$CO$_3$. The hydrogenolysis of glycerol was easier on bimetallic Pd-Cu/solid-base catalysts than over separated Pd and Cu catalysts. On performing the hydrogenolysis of glycerol on Pd$_{0.04}$Cu$_{0.4}$/Mg$_{5.5}$Al$_2$O$_{8.5}$, at 180 °C for 10 h and 20 bar of H$_2$, both in ethanol and methanol solutions, conversion and selectivity to 1,2-PDO are particularly efficient [52]. Authors suggest that the performances are better than in water, and stem from the minor interaction between ethanol and the catalyst surface making more surface available for the conversion of glycerol [52,53].

Table 2. Hydrogenolysis of glycerol in presence of H-donor solvents.

Entry	Catalyst	Solvent[1]	Cat/Gly[2]	Temp (°C)	Gas (bar)	Time (h)	Conv. (%)	Desired Prod. Select. (%)[3]	Ref.
1	Pd$_{0.04}$Cu$_{0.4}$/Mg$_{5.5}$Al$_2$O$_{8.5}$	MeOH	0.125	180	H$_2$ (20)	10	89.5	1,2-PDO (98)	[52]
2	Pd$_{0.04}$Cu$_{0.4}$/Mg$_{5.5}$Al$_2$O$_{8.5}$	EtOH	0.125	180	H$_2$ (20)	10	88.0	1,2-PDO (99)	[52]
3	Rh$_{0.02}$Cu$_{0.4}$/Mg$_{5.6}$Al$_{1.9}$O$_{8.6}$	EtOH	0.167	180	H$_2$ (20)	10	91.0	1,2-PDO (99)	[54]
4	Pd/Fe$_3$O$_4$	2-PO	0.237	180	H$_2$ (5)	24	100	1,2-PDO (71)	[55]
5	2Pt/20WO$_3$/ZrO$_2$	EtOH	0.250	170	H$_2$ (55)	12	45.7	1,3-PDO (21)	[56]

[1] Abbrevations: EtOH: ethanol; 2-PO: 2-propanol; [2] Ratio of Cat/Gly (g/g); [3] Abbreviations: 1,2-PDO: 1,2-propanediol; 1,3-PDO: 1,3-propanediol.

On using analogous catalysts based on Rh (Rh$_{0.02}$Cu$_{0.4}$/Mg$_{5.6}$Al$_{1.98}$O$_{8.6}$), the hydrogenolysis of glycerol, in the presence of ethanol, leads to high conversion and selectivity to 1,2-PDO respectively, 91.0% and 98.7%, at 2.0 MPa H$_2$ and 180 °C. Moreover, this catalyst was found to be stable for five consecutive hydrogenolysis tests in ethanol, even if the conversion decreases from 91% to 56.7% in the third cycle and then it remains constant until the fifth cycle [54]. Similarly, for the Pd-Cu/solid-base catalyst, the improved performance was attributed to a less strong interaction of the solvent with the catalytic surface [54].

Also, the bimetallic Pd/Fe$_3$O$_4$ catalyst was tested for the hydrogenolysis of glycerol using 2-propanol, as a solvent, and in mild operating conditions, such as 180 °C and only 5 bar of molecular hydrogen [55].

Gong et al. have found that it is possible to guide the selectivity of glycerol towards 1,3-propanediol using a 2Pt/20WO$_3$/ZrO$_2$ catalyst, in ethanol as solvent medium [56].

Another interesting route to valorize glycerol, like a potential biorefinery feedstock, is to obtain allyl alcohol by using H-donor molecules as a catalyst and/or solvent, through the dehydration of glycerol to acrolein followed by reductive H-transfer to allyl alcohol (Scheme 3).

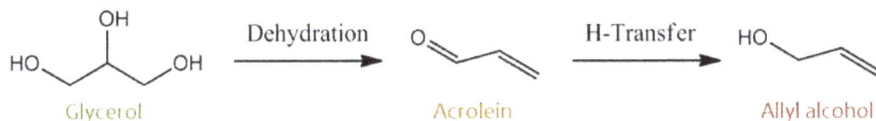

Scheme 3. Schematic representation of the dehydration/H-transfer of glycerol into allyl alcohol.

Schüth and coworkers investigated the conversion of glycerol into allyl alcohol, through an initial dehydration to acrolein, by using iron oxide as a catalyst [57]. Among the operating conditions investigated, it was found that at 320 °C, it is possible to obtain an almost full conversion of glycerol and a yield to allyl alcohol of 20–25%. Particularly, the selectivity in the transfer hydrogenation to allyl alcohol, is close to 100%. This evidence has never been observed before for iron oxide catalysts, which were scarcely considered active in hydrogen transfer reactions. In this case the hydrogen donor species could be the glycerol itself and some intermediates bearing hydroxy groups [57].

Furthermore, Masuda and coworkers carried out the conversion of glycerol into allyl alcohol using iron oxide-based catalysts at 350 °C [58]. The dehydration of glycerol takes place on acid sites of catalysts, while the allyl alcohol formation occurs through a hydrogen transfer mechanism. Several alkali metals (Na, K, Rb, and Cs) were supported on the ZrO$_2$–FeO$_x$ substrate and all of them give higher allyl alcohol yield suppressing glycerol dehydration due to the reduced catalyst acidic property. Particularly, the K-supported catalyst (K/ZrO$_2$–FeO$_x$) affords an allyl alcohol yield of 27 mol%. Also, in this case, the hydrogen transfer mechanism seems to take place between glycerol and hydrogen atoms deriving from the formic acid formed during the reaction, or active hydrogen species produced from the decomposition of H$_2$O by ZrO$_2$. Furthermore, the addition of Al$_2$O$_3$ (K/Al$_2$O$_3$–ZrO$_2$–FeO$_x$) enables an improvement of the stability of the catalyst during the glycerol conversion, also making the K/Al$_2$O$_3$–ZrO$_2$–FeO$_x$ catalyst applicable directly to the crude glycerol (which is the waste solution obtained from biodiesel production), reaching a yield of 29% in allyl alcohol after 4–6 h of reaction [58].

Another kind of approach was followed by Bergman and coworkers, by using formic acid either as acid catalyst or solvent, with the aim to deoxygenate the glycerol into an allyl compound through the mediation of formic acid (230–240 °C) [59]. This method allows for the conversion of the 1,2-dihydroxy group to a carbon–carbon double bond. The same procedure was applied to erythritol that has been converted into 2,5-dihydrofuran at 210–220 °C [59]. During the dehydration step, formic acid acts as acid catalyst and, in the reductive step, as hydrogen donor [60].

A similar approach was followed by Fristrup and co-workers and allows for the reduction of two vicinal diols into an alkene group [61]. In this work the deoxydehydration (DODH) of glycerol and erythritol is performed with the commercially available (NH$_4$)$_6$Mo$_7$O$_{24}$·4H$_2$O catalyst, in the presence of 2-propanol, acting both as solvent and reducent agent. Following this approach, the total yield of reduced species (such as alkene and alcohols) can be as high as 92% at 240–250 °C. The DODH of erythritol can reach a yield of 39% to 2,5-dihydrofuran [60].

2.2. Glucose and Carbohydrates

Deng et al. have investigated the conversion of the biomass derived cellulose, starch, and glucose into γ-valerolactone without using any external hydrogen source. The first step implies the dehydration of biomass carbohydrates into levulinic and formic acids, whereas in the second step the formic acid furnishes the hydrogen necessary to the reduction of levulinic acid into γ-valerolactone [61]. The authors hydrolyzed carbohydrates (microcrystalline cellulose, α-cellulose and starch) using a solution of 0.8 M HCl at 220 °C in order to obtain levulinic acid and formic acid. In this step it is very important that the yield to formic acid has to be high enough (in excess or in equimolar amount with respect to levulinic acid) to enable the subsequent reduction to levulinic acid. In this work, the recyclable and cheap $RuCl_3$/PPh_3/pyridine catalyst was used. In a model experiment, performed using glucose, γ-valerolactone was produced with a yield of 48% [61].

Another example of combined dehydration and transfer-hydrogenation to produce γ-valerolactone, starting from glucose or fructose, was given by Heeres [62]. In this case the process was performed in water using the trifluoroacetic acid (TFA) coupled with a heterogeneous hydrogenation catalyst (Ru/C), and also molecular hydrogen or formic acid as hydrogen donor were added [62].

Au-based catalysts were investigated by Fan and co-workers [63] and the Au/ZrO_2 catalyst shows interesting results in converting efficiently glucose into levulinic and formic acids in high yield, 54% and 58% respectively. Under the same reaction conditions also cellulose, starch, and fructose were converted [63].

Scholz et al. applied co-precipitated Cu-Ni-Al catalysts to the hydrogenation of glucose by using 1,4-butanediol as a hydrogen source to obtain sorbitol. A sorbitol yield of 67% was obtained from glucose with the catalyst remaining stable within 48 h of reaction. Similarly, several other substrates (i.e., fructose, mannose, xylose, arabinose) can be used to obtain the corresponding polyols [64].

Finally, there is also an example, given by Van Hengstum et al., where glucose is applied as H-donor substrate [65]. An equimolar mixture of glucose and fructose was employed to obtain gluconic acid and hexitols, such as sorbitol and mannitol. On using Pt/C and Rh/C catalysts, it is possible to obtain equal amounts of gluconic acid and hexitols, operating at room temperature in an aqueous alkaline medium, under nitrogen atmosphere. The general mechanism that occurs in this reaction starts from the generation of hydrogen from the dehydrogenation of glucose; the hydrogen generated is chemisorbed on the metallic surface of the catalyst and it is subsequently consumed by the co-adsorbed fructose [65].

2.3. Cellulose

At present, the work of Fukuoka and coworkers is the only example of the catalytic transfer hydrogenation applied to cellulose to obtain hexitols, such as sorbitol and mannitol, without using high molecular hydrogen pressures, but only hydrogen producing in situ molecules, such as 2-propanol (Table 3). A screening of several Ru-based catalysts was carried out by using milled cellulose in aqueous solution at 25 vol% of 2-propanol at 190 °C for 18 h [66] showing that the support plays a crucial role for the reactivity of the catalysts. Indeed, supported Ru/carbons appears the more reactive, and in particular Ru/C-Q10, Ru/CMK-3 and Ru/AC(N) show high conversion (74–81%) and high yields to sorbitol and mannitol (sum of C6-polyols 42.5–45.8%). Other supports, such as Al_2O_3, TiO_2, and ZrO_2, were found inactive. Characterization measurements, show also that the reactivity stems from the presence of highly dispersed cationic ruthenium species, which are active for the transfer hydrogenolysis [66].

Beltramini, in collaboration with Fukuoka, carried out further studies using catalysts of Ru supported on activated carbon applied to the CTH of cellulose, with the scope of optimizing the operating conditions for reactions performed in batch mode, overcoming the problem deriving from the long reaction time, in order to carry on the process with a continuous set-up in a fixed bed reactor [67]. In particular, glucose gives 82% of conversion and a 79.7% yield in hexitols at 180 °C after only 20 min of reaction, using water and 2-propanol in equal volume. The optimized conditions were subsequently applied to the transfer hydrogenation of cellulose oligomers, which were obtained through a process of milling of crystalline cellulose impregnated with sulfuric acid. This pre-treatment was necessary in order to facilitate the solubilization of the reacting substrate, used in a continuous process, and to reduce the time of reaction. In this way, within 20 min of reaction at 180 °C, the highest yield of 35.3% to hexitols, was obtained. Furthermore, on performing the reaction in a continuous set-up, a hexitols yield of 36.4%, constant for 12 h of reaction, was obtained with a liquid hourly space velocity (LHSV) of 4.7 h^{-1}. The mechanism that enables the transfer of hydrogen from 2-propanol to glucose seems to be the di-hydride mechanism [67].

Table 3. Catalytic transfer hydrogenolysis of cellulose performed in batch conditions.

Entry	Substrate [1]	Catalyst	H-Donor [2]	Temp (°C)	Time (h)	Conv. (%)	Yield Sorbitol (%)	Yield Mannitol (%)	Ref.
1	MC	Ru/C-Q10	2-PO	190	18	80.2	36.8	9.0	[66]
2	MC	Ru/CMK-3	2-PO	190	18	81.2	35.7	9.3	[66]
3	MC	Ru/AC(N)	2-PO	190	18	74.4	33.5	9.0	[66]
4	Glucose	Ru/AC(N)	2-PO	180	0.33	82	77.0	2.7	[67]
5	ACO	Ru/AC(N)	2-PO	180	0.33	100	32.2	3.1	[67]

[1] Abbrevations: MC: milled cellulose; ACO: acidified cellulose oligomers; [2] Abbreviations: 2-PO: 2-propanol.

3. Catalytic Transfer Hydrogenolysis (CTH) Reactions of Hemicellulose Derived Molecules

3.1. Furfural Derivatives

Furfural (FU) is an important building block for biorefineries and chemical industries and it became a subject of interest for both academic and industrial research [68,69]. It is commercially produced by the acid-catalysed reaction of biomass containing pentose sugars [70], and it is the starting molecule for several bio-based chemical intermediates (Figure 2).

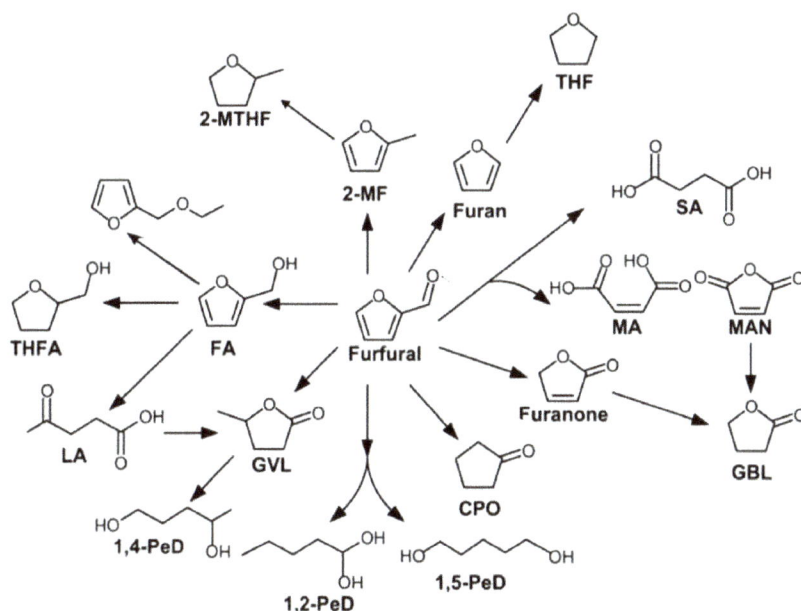

Figure 2. Examples of furfural-derived chemicals and biofuels. Adapted from Ref. [70]. Copyright Year 2016, American Chemical Society.

Among all possible reaction pathways, the hydrogenation/hydrogenolysis of FU to 2-methylfuran (MF) and 2-methyltetrahydrofuran (MTHF) gained much attention recently because of their fuel properties [70–72]. Many works were also directed to the use of alcohols as a H-source in CTH reactions of FU and an overview referring to heterogeneous metal catalysts used in the CTH of FU is presented in Table 4.

Table 4. A literature overview of the examples of metal catalysts used in CTH of FU.

Entry	Catalyst	H-Donor [1]	Reaction Conditions [2] (Temperature, Time, Solvent)	Conv. (%)	Main Product [3]	Yield (%)	Ref.
1	Ru/RuO$_2$/C	2-PE, 2-BU	180 °C, 10 h, 2-PE, 2-BU	100.0	MF	76.0	[73]
2	Ru/RuO$_x$/C	2-BU	180 °C, 10 h, TU	100.0	MF	76.0	[74]
3	Ru/C	2-PO	180 °C, 10 h, 2-PO	100.0	MF	61.0	[75]
4	Ru/NiFe$_2$O$_4$	2-PO	180 °C, 10 h, 2-PO	>97.0	MF	83.0	[76]
5	Cu-Ni/Al$_2$O$_3$	2-PO	230 °C, 4 h, 2-PO	>97.0	MF, MTHF	82.5	[77]
6	Cu/C	2-PO	200 °C, 5 h, 2-PO	96.3	MF	84.0	[78]
7	Cu-Pd/C	2-PO	200 °C, 4 h, 2-PO	100.0	MF, MTHF	83.9	[79]
8	Cu$_3$Al-A	MeOH	240 °C, 1.5 h, MeOH	>97.7	MF	88.2	[80]
9	Pd/Fe$_2$O$_3$	2-PO	180 °C, 7.5 h, 2-PO	95.0	MF, MTHF	62.0	[81]

[1] Abbreviations: 2-PE: 2-pentanol; 2-BU: 2-butanol; 2-PO: 2-propanol; MeOH: 2-methanol; [2] Abbreviations: TU: toluene; [3] Abbreviations: MF: 2-methylfuran; MTHF: 2-methyltetrahydrofuran.

Vlacos and co-workers deeply investigated the CTH of FU [73–75]. As an example, they showed the effect of alcohols, both as a solvent and hydrogen donor, in the CTH of FU to MF over the Ru/RuO$_2$/C catalyst [73]. The correlation between the type of alcohol and the yield to MF was proposed. MF yield increases from 0 to 68% at 180 °C according to the following order: 2-methyl-2-butanol < tert-butanol < ethanol < 1-propanol ≈ 2-propanol < 2-butanol ≈ 2-pentanol. The highest yield was obtained on using 2-butanol and 2-pentanol.

Liang research group presented a very stable Ru/NiFe$_2$O$_4$ catalyst for the CTH of FU into MF in high yield (83%) at relative mild conditions (180 °C, 10 h, 2.1 MPa N$_2$) by using 2-PO as H-donor [76].

Zhang et al. reported the CTH of FU with 2-propanol used as a hydrogen donor. The Cu-Ni/Al$_2$O$_3$ bimetallic catalyst, synthesized using the coprecipitation method, shows an improved activity in the mixed production of MF and MTHF compared to monometallic catalysts, with a yield of 82.5% [77].

Gong et al. [78] reported on the catalytic results related to a Cu/C catalyst with different Cu loadings (10.4; 17.1; 22.9 wt%), prepared by the ultrasound-assisted impregnation method, performed in the CTH of FU. The highest conversion of FU and the selectivity to MF were observed with a Cu loading of 17.1 wt% (96.3% FU conversion and 58.8% MF selectivity) and 2-propanol as hydrogen source. The effect of the hydrogen donor ability in the on CTH of FU to MF was also investigated. The selectivity to MF increases from 8.9% to 78.5% following the trend: 2-propanol > 2-pentanol > 2-butanol > ethanol > methanol > n-butanol > n-pentanol (180 °C, 5 h). The highest selectivity of 91.6% to MF was obtained in presence of 2-propanol and was completed within 5 h at 200 °C. This effect was attributed to the dispersing effect of the support that prevents aggregation of Cu nanoparticles.

The CTH of FU into 2-MF and 2-MTHF, promoted by several bimetallic catalysts in presence of 2-propanol as H-donor, was investigated by Huang [79]. The best performances were obtained with the bimetallic Cu-Pd catalyst. Moreover, authors demonstrated that selectivity toward 2-MF or 2-MTHF can be driven by changing the Pd ratios in the Cu-Pd system.

Zhang and Chen [80] investigated a series of copper-based catalysts Cu$_x$Al-A (where x refers to the Cu/Al molar ratio and -A means that the catalyst was activated in H$_2$/N$_2$ flow) obtained from hydrotalcite precursors in the CTH of HMF to 2,5-dimethylfuran (DMF) with methanol both as a solvent and hydrogen source. The Cu$_3$Al-A catalysts showed the best catalytic activity and good recycling performances were also observed.

Scholz and co-workers [81] used monometallic Cu/Fe$_2$O$_3$, Ni/Fe$_2$O$_3$ and Pd/Fe$_2$O$_3$ with different metal loading (1–10 wt%), synthesised by the coprecipitation method, in the CTH of FU in presence of 2-propanol. The hydrogenolysis selectivity highly depends both on the metal nature and its loading. The highest activity was observed with the 2 wt% Pd/Fe_2O_3 and was attributed to a strong metal-support interaction. A significant enhancement of the MF and MTHF yield to 62.0% was observed under continuous flow conditions.

Yang et al. compared the CTH of FU using three different hydrogen sources: 2-propanol, formic acid (FA), and molecular hydrogen in the presence of a 2%Ni-20%Co/C catalyst [82]. It has been observed that with 2-propanol, the conversion of FU was 51% at 190 °C (24 h) with the main product being furfuryl alcohol (FUA) with traces of MF. Increasing of the temperature to 210 °C causes the increasing of the conversion to 99%, but the yield to MF remains low, about 5%. When formic acid was used as an H-donor (210 °C, 24 h) the conversion was 99% and a strong increase of the yield to MF up to 79.2% was observed. This result is even better if compared to that obtained by using molecular hydrogen (33.2% yield of MF at 210 °C and 1.5 MPa for 24 h). Authors propose a relative reaction pathway (Scheme 4) [82].

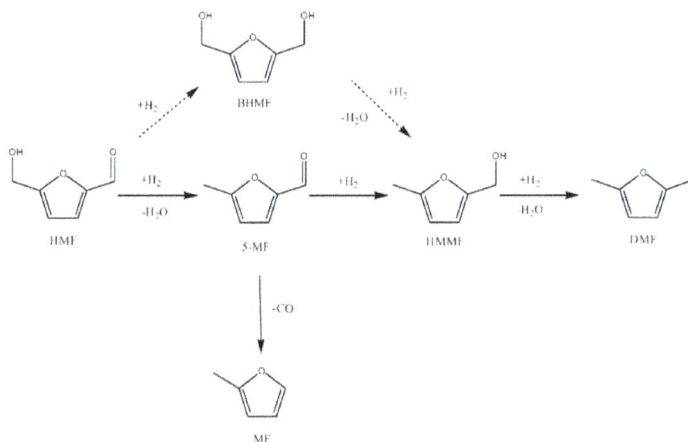

Scheme 4. Proposed reaction pathway for the CTH of HMF. Adapted from Ref. [82]. Copyright Year 2017, Elsevier.

5-hydroxymethylfurfural (HMF) can be synthesized by dehydration of hemicellulose and cellulose and it is one of the most promising feedstocks because of its abundance. In the biorefinery area, HMF is regarded as a "sleeping giant" in the field of value-added and renewable resources [83]. HMF possesses, in fact, two functional groups and can be converted to valuable compounds useful in chemical manufacturing and industrial applications.

Jae and co-workers [84] investigated the catalytic performance of a 5% Ru/C catalyst in the presence of 2-propanol. It was shown that the selectivity towards MF increases with increasing reaction temperatures. At low temperatures (100–130 °C) the primary product is 2,5-bis(hydroxymethyl)furan (BHMF) after 5 h of reaction. On increasing the temperature to 190 °C, BHMF is completely converted into MF with a selectivity up to 81%.

Wang et al. [85] used Ru/Co_3O_4 catalysts (prepared via the co-precipitation method) in the CTH of HMF to BHMF. It was found that the effect of reaction temperature and time is critical and a simplified reaction pathway from HMF to BHMF was proposed (Scheme 5).

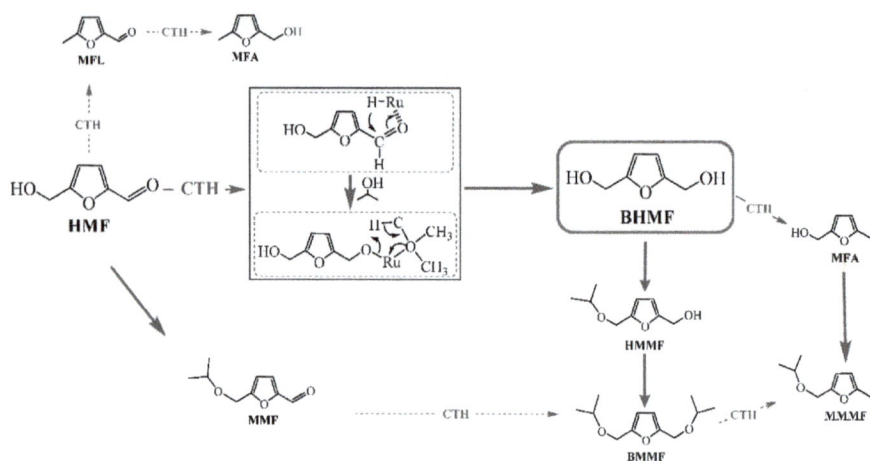

Scheme 5. Simplified reaction pathways in the CTH of HMF to BHMF using 2-propanol and the Ru/Co$_3$O$_4$ catalyst. Adapted from Ref. [86].

Aelig et al. tested some Cu/AlO$_x$ catalysts in the CTH of HMF to BHMF in presence of 1,4-butanediol as an H-source and 1,4-dioxane as a solvent in continuous flow reactions [86]. It has been observed that the highest conversion of HMF with a BHMF selectivity of 93% was obtained at 220 °C.

Hansen et al. [87] reported the catalytic performance of Cu-PMO (porous metal oxide) catalysts in the CTH of HMF to DMF with the supercritical methanol. The highest conversion (>99%) was obtained at 320 °C with a selectivity of 32% to DMF (3 h reaction time).

The CTH of HMF to BHMF and the following etherification to BMF were investigated by Jae et al. [88]. Reactions were catalysed by Lewis acid zeolites, Zr-Beta and Sr-Beta in the presence of 2-propanol (170 °C, 6 h). More than 80% of yield towards 5-bis(alkoxymethyl)furan (BMF) was obtained. Catalytic results showed that the etherification of HMF and BHMF with the alcohol is the fast reaction and the Meerwein–Ponndorf–Verley conversion of HMF to BHMF is the rate-determining step.

Hao et al. [89] reported that the low-cost ZrO(OH)$_2$ was effective in CTH of HMF to BHMF in the presence of ethanol (the HMF conversion of 94%, and an almost 89% selectivity to BHMF was obtained at 150 °C in 2.5 h).

A CTH transformation of HMF to BHMF over various magnetic zirconium hydroxides (MZH) in the presence of 2-butanol as a hydrogen source was also recently reported by Hu et al. [90]. An excellent catalytic activity was observed over MZH with Zr/Fe = 2 molar ratio, resulting in 98.4% HMF conversion and 89.6% DHMF yield at 150 °C for 5 h. A reaction mechanism for the CTH of HMF to DHMF was also proposed where the hydroxyl groups, with the aid of zirconium metal centres, were responsible for the hydride transfer via a ring structure.

Methanol as a clean and useful H-source was reported by Prof. Cavani's research group in the CTH of HMF to BHMF in the presence of MgO as a catalyst [91]. A superior 100% conversion of HMF and 100% selectivity to BHMF at 160 °C in 3 h was observed and the only methanol deriving products were CO, CO$_2$, and CH$_4$, whereas partially hydrogenated or dimerized compounds, such as formaldehydes, hemiacetals, and acetals, were not detected. Recovery experiments also showed that MgO can be recovered by filtration and re-used, but a thermal treatment is required to regenerate the partially deactivated catalyst.

Studies with FA as an H-source in the CTH of HMF were carried out to a considerably lesser extent compared to that of alcohols. In the reaction of HMF with the participation of FA as a hydrogen source (the reaction was carried out in an organic solvent (THF) with the addition of H$_2$SO$_4$ in the presence of Pd/C at 80 °C), Thananatthanachon and Rauchfuss obtained a DMF yield of 94% [92]. The disadvantage of this approach is the formation of esters as by-products. The same authors also studied some homogeneous noble metal catalysts in the CTH with FA as hydrogen source and THF as solvent [93].

Tuteja and co-workers [94] reported the CTH process of HMF to 1,6-hexanediol (HDO) under atmospheric pressure in the presence of Pd/ZrP (zirconium phosphate) with FA as a hydrogen source. A 43% yield of HDO at 140 °C in 21 h was obtained. This effect was correlated with the specific Brønsted acidity of the ZrP support.

Gao et al. report a high DMF and 2,5-dimethyltetrahydrofuran (DMTHF) yield of 96.1% and 94.6%, respectively by using a nitrogen carbon doped-Cu/MgAlO catalyst in the CTH of HMF with cyclohexanol as a hydrogen source (220 °C, 0.5 h) and suggest that highly dispersed Cu^0 nanoparticles and electrophilic Cu^+ species promote the hydrogen transfer and activation of both the carbonyl and hydroxyl groups in the HMF molecule [95].

3.2. Levulinic Acid

The upgrading of biomass-derived levulinic acid (LA) and its esters to γ-valerolactone (GVL) is a very important issue in the development of a sustainable and economical route to chemicals and liquid fuels. LA has attracted much attention because it is one of the top biomass derived platform molecules that can be made by the transformation of the lignocellulosic biomass followed by hydration and dehydration of hexose sugars [96,97]. The different chemical products that can be derived from LA are collected in Figure 3.

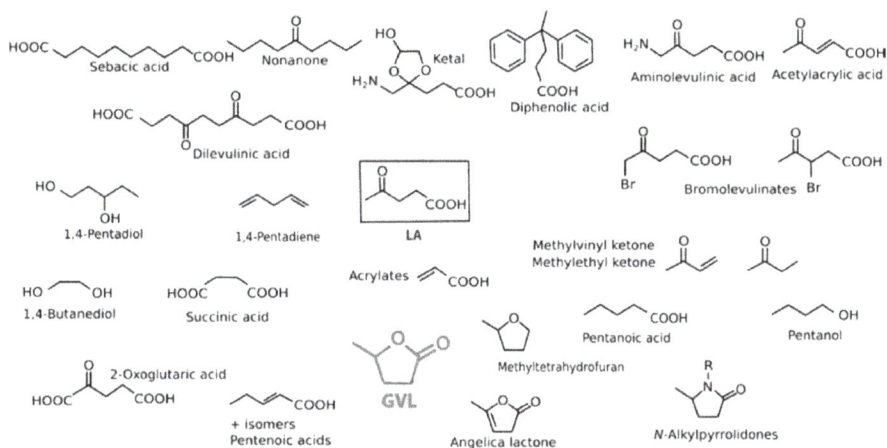

Figure 3. Chemical products derived from LA. Adapted from Ref. [98].

The CTH of LA can be an alternative way to the typical reduction of the molecule to GVL by means of molecular hydrogen and an increasing interest for this type of reactions is growing in the literature. Some of the most representative heterogeneous catalysts for the CTH of LA in the presence of alcohols as an H-source are collected in Table 5.

The CTH of LA promoted by heterogeneous Ni catalysts (Al_2O_3, ZnO, MMT and SiO_2) was reported by Rode and co-workers. GVL was obtained in very high yield (~99%) over the Ni/MMT catalyst within 1 h [98].

Yang et al. referred to a CTH process for the production of GVL from EL and 2-propanol as an H-donor [99]. The process was performed at room temperature over a RANEY® Ni catalyst with a yield of GVL of 99.0%.

Song and co-workers studied some porous Zr-containing catalysts bearing a phenate group in the CTH of EL to GVL [100]. The results show that Zr-HBA is very active for the conversion of EL and the GVL yield of 95.9% can be achieved in the presence of 2-butanol as an H-donor.

Tang et al. synthesised low cost and eco-friendly metal hydroxides and examined their behaviour as catalysts in the production of GVL from biomass-derived EL via CTH in the presence of 2-propanol [101,102]. A 93.6% conversion of EL and 94.5% yield to GVL was achieved at 250 °C in 1 h with 2-propanol.

Table 5. Representative heterogeneous catalysts in the CTH of LA and its derivatives using different alcohols as hydrogen donors.

Entry	Substrate [1]	Catalyst [2]	H-Donor [3]	Reaction Conditions (Temperature, Time, Solvent)	Conv. (%)	GVL Yield (%)	Ref.
1	LA	Ni/MMT	2-PO	200 °C, 1 h, 2-PO	99.0	99.0	[98]
2	EL	Raney® Ni	2-PO	25 °C, 9 h, 2-PO	-	99.0	[99]
3	EL	Zr-HBA	2-BU	150 °C, 4 h, 2-BU	100.0	95.9	[100]
4	EL	ZrO_2	EtOH	250 °C, 3 h, EtOH	95.5	81.5	[101]
5	EL	$ZrO(OH)_2$	2-PO	200 °C, 1 h, 2-PO	93.6	94.5	[102]
6	BL	ZrPO-1.00	2-PO	210 °C, 2 h, 2-PO	98.1	95.7	[103]
7	ML	ZrO_2/SBA-15	2-PO	150 °C, 6 h, 2-PO	99.9	95.0	[104]
8	LA	Zr-Beta	2-PO	250 °C, vap. phase, 2-PO	100.0	>99.0	[105]
9	LA	ZrO_2	2-BU	150 °C, 16 h, 2-BU	>99.9	84.7	[106]
10	FU	Zr-Beta + Al-MFI-ns	2-BU	120 °C, 48 h, 2-BU	-	78.0	[107]
11	EL	UiO66(Zr)	2-PO	200 °C, 2 h, 2-PO	>98.0	92.7	[108]

[1] Abbreviations: LA: levulinic acid; EL: methyl levulinate; FU: furfural; BL: n-butyl levulinate; ML: methyl levulinate; [2] Abbreviations: MMT: montmorillonite; [3] Abbreviations: 2-PO: 2-propanol; 1,4-BU: 1,4-butadienol; 2-BU: 2-butanol EtOH: ethanol.

Li studied the efficient production of GVL from LA and its esters via the CTH using different hydrogen donors [103]. Zirconium phosphate catalysts (ZrPO-X, where X is the molar ratio of Zr to PO) were tested and a 98.1% of BL conversion and a 95.7% GVL yield were obtained using ZrPO-1.00 system.

Kuwahara and co-workers investigated a series of ZrO_2 catalysts supported on SBA-15 silica in the synthesis of GVL from LA and its esters via CTH using different alcohols as hydrogen donors [104]. The highest yield of GVL (95.0%) was obtained from methyl levulinate (ML) in the presence of 2-propanol (150 °C, 6 h, conv. 99.9%). A reaction mechanism of the CTH was also proposed (Scheme 6).

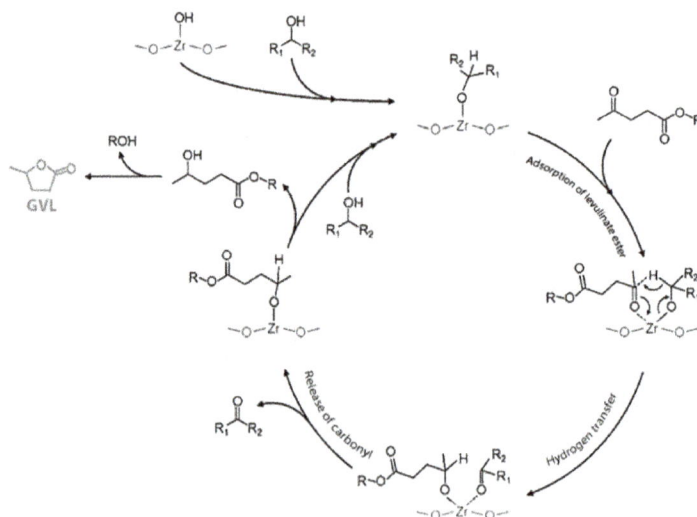

Scheme 6. Proposed reaction mechanism of the CTH of levulinate esters over silica-supported ZrO_2 catalysts. Adapted from Ref. [106]. Copyright Year 2017, Elsevier.

Accordingly, Zr–Beta zeolite was proved to be a very efficient catalyst in the CTH of LA to GVL via Meerwein–Ponndorf–Verley reduction [105].

Chia et al. examined different metal oxides (MgO/Al_2O_3, MgO/ZrO_2, ZrO_2) in the presence of different alcohols used as hydrogen donors [106]. ZrO_2 turned out to be the most active and selective catalyst when 2-butanol was used as a hydrogen source.

Bui reported the one-pot conversion of FU to GVL through the CTH process catalysed by zeolites containing Lewis and Brønsted acid sites (Zr-Beta + Al-MFI-ns). The best GVL yield was 78.0% at 120 °C in 48 h using 2-butanol as a hydrogen source [107].

Valekar and co-workers studied zirconium-based metal organic frameworks catalysts for the CTH of EL to GVL in the presence of 2-propanol: a yield of 98.7% to GVL with the conversion of EL almost of 100.0% at 200 °C within 2 h was reported [108].

Taking into account that formic acid is co-produced along with LA during the biomass conversion, it seems very interesting to also use this molecule in CTH reactions of LA and its esters [109]. Catalytic systems for the CTH of LA and its esters to GVL using FA as an H-source are summarised in Table 6.

Table 6. Catalytic systems for the production GVL from LA and derivatives using FA as a hydrogen source.

Entry	Catalyst	Reaction Conditions (Temperature, Time, Solvent) [1]	Conv. (%)	GVL Yield (%)	Ref.
1	Ru NPs	130 °C, 42 h, FA + triethylamine + water	100.0	100.0	[110]
2	Ru-P/SiO$_2$ + Ru/TiO$_2$	150 °C, 6 h, LA	100.0	30.0	[111]
3	Ru/C	150 °C, 5 h, water	100.0	90.0	[112]
4	Cu/SiO$_2$	250 °C, vap. phase, -, FA + water	48.0	90.0	[113]
5	Cu/ZrO$_2$	200 °C, 5 h, water	100.0	100.0	[114]
6	Ag-Ni/ZrO$_2$	220 °C, 5 h, water	100.0	99.0	[115]

[1] Abbreviations: FA: formic acid.

Ortiz-Cervantes and García synthesised free Ru nanoparticles from the Ru_3Co_{12} for the CTH process of LA to GVL in presence of formic acid [110]. A complete conversion of LA and 100.0% yield to GVL were obtained at 130 °C in 42 h.

Deng and co-workers used immobilised Ru-P/SiO$_2$ + Ru/TiO$_2$ to carry out the reaction of LA to GVL showing that the two-step process implies the decomposition of FA and the CTH of LA through parallel routes [111].

Son et al. reported a 90.0% GVL yield in the CTH process using FA as hydrogen source and Ru/C in optimised reaction conditions (150 °C, 5 h) [112].

Cu-supported catalysts were also tested in the CTH of the LA conversion to GVL by Lomate et al. [113]. Results indicate that Cu/SiO$_2$ leads to a conversion of 48.0% and a 90.0% yield to GVL. Additionally, Cu/SiO$_2$ shows a remarkable stability and re-utilization that indicates minimal loses of Cu particles during the reaction.

Yuan et al. described the successful utilization of Cu/ZrO$_2$ in the CTH process of LA to GVL by optimising the reaction conditions (200 °C, 5 h) in order to get a complete conversion of LA and a 100.0% GVL yield [114].

Finally, Hengne reported the total conversion of LA to GVL in the CTH process in the presence of (10% Ag, 20% Ni) Ag-Ni/ZrO$_2$. The synergistic effect between Ag and Ni is responsible of the decomposition of FA and the in-situ hydrogenation of LA precedes smoothly to GVL with a 99.0% yield [115].

4. Catalytic Transfer Hydrogenolysis (CTH) of Lignin and Its Derived Molecules

4.1. CTH of Lignin Derived Molecules

Prof. Ford's research group can be surely considered one the first that studied the transfer hydrogenolysis of dihydrobenzofuran (DHBF), a lignin model compund (α-O-4 C-O bond), using the Cu-doped PMO as a catalyst and MeOH as a hydrogen source/solvent, in a microreactor, for 2 h at 300 °C, in the presence of KOH. DHBF was fully converted to methylated 2-ethylphenols (63%), 2-ethylphenol (22%) and phenol (11%) [116].

Besse and co-workers investigated the catalytic transfer hydrogenolysis of eight model compunds with peculiar lignin linkages at 275–350 °C in a batch reactor, using the Pt/C catalyst and EtOH/H2O as a hydrogen source solvent. They demonstrated that the lignin linkage cleavage follows the energy bond order showing that methoxyl and phenolic hydroxyl model molecules are unreactive, while α-1 model compounds are fully converted [117].

Han and co-workers recently demonstrated the performing catalytic activity of the commercial Ru/C in CTH reactions of aromatic ethers using a variety of 4-O-5 type lignin model compounds using 2-propanol as a solvent/H-donor under mild conditions (at 120 °C for 10–26 h) [118].

Samec and co-workers report that the commercial Pd/C can be a good catalyst in the C–O bond cleavage of the β-O-4′ ether as a model lignin molecule using formic acid and 2-propanol as H-donors. They also tested other heterogeneous catalysts (Ir/C, Ni/C, Pd/C, Re/C, Rh/C) and Pd/C showed the higher reactivity in the cleavage of the β-O-4′ C-O bond, allowing an efficient transformation to the corresponding aryl ketones and phenols in high yield, at 80 °C for 1–24 h. They proposed a reaction mechanism in which the first key step is the dehydrogenation of the α-CHO group followed by formation of a Pd-enolate complex, that undergoes a transfer hydrogenolysis process [119,120].

Wang et al., encouraged by the catalytic performance of Pd catalysts under hydrogenation conditions, studied the CTH of phenol in the presence of formic acid under mild reaction conditions: a good selectivity of 80% to cyclohexanone was observed [121].

In 2012, Rinaldi and co-workers reported, for the first time, the use of the bimetallic RANEY® Ni catalyst in the H-transfer reactions of lignin model molecules. 2-propanol was used as reaction solvent and hydrogen source and 32 model substrates at temperatures from 80 °C to 120 °C for 3 h were explored: the RANEY® Ni catalyst shows a high performance under CTH conditions and a good stability in the recycling tests [122]. In the course of years, his research group deeply investigated the CTH of other lignin model molecules (including phenol) in the presence of different heterogeneous catalysts, also elucidating the role of the catalyst's surface on the H-transfer mechanism [123,124].

Recently, a Pd/Ni catalyst was found to be very efficient also in the CTH of other model molecules, such as diphenyl ether (DPE), 2-phenethyl phenyl ether (PPE), and benzyl phenyl ether (BPE), leading to an arene derivative [125]. Interestingly, authors found that the hydrogenation of an aromatic ring in the CTH of DPE, PPE, and DPE is influenced by the nature of aryl groups that compose the aromatic ether (Figure 4).

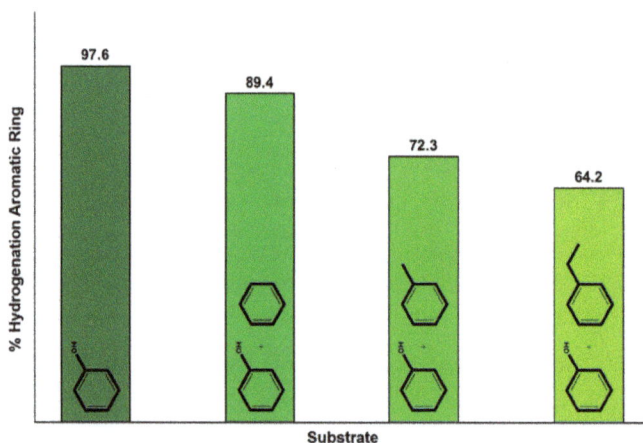

Figure 4. Competitive catalytic hydrogenation of phenol and phenol/benzene, phenol/toluene and phenol/ethylbenzene equimolar mixtures at 210 °C under CTH conditions. Adapted from Ref. [125]. Copyright Year 2018, American Chemical Society.

The same authors previously presented the bimetallic Pd/Fe_3O_4 system as a very efficient catalyst in the cleavage of the C–O bond in aromatic ethers and esters for the production of arene derivatives in presence of 2-propanol as a solvent/H-donor [126,127]. In the case of benzyl phenyl ether (BPE), authors proposed that the CTH process is very sensitive to the steric hindrance of the involved H-donor system, postulating a reaction mechanism in which the H-transfer from the alcoholic solvent and the C–O bond cleavage occur in a single step (Scheme 7).

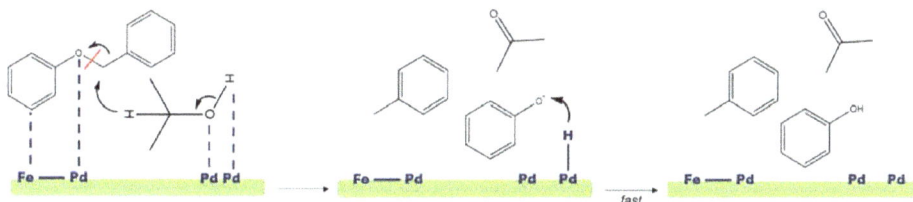

Scheme 7. CTH mechanism of BPE promoted by Pd/Fe_3O_4 catalyst in presence of 2-propanol as H-donor/solvent. Adapted from Ref. [127]. Copyright Year 2016, Royal Society of Chemistry

Kim and co-workers explored the CTH of guaiacol, another important lignin model compound, to ciclohexane (>70% yield) using the bimetallic RuRe/C catalytic system and 2-propanol as a hydrogen source/solvent, demonstrating that the bimetallic catalyst is very efficient both in the dehydrogenation of solvent and the hydrogenation of guaiacol, and that the presence of Re increases the rate of the C-O hydrogenolysis, allowing a higher selectivity to cyclohexane (≈60%) [128].

Wang and co-workers proposed a strategy for the direct deoxygenation of *p*-cresol to toluene (84% yield) via the catalytic transfer hydrogenolysis, promoted by the Ru/Nb_2O_5-SiO_2 catalyst using 2-propanol as a hydrogen source/solvent at 230 °C. Authors demonstrated the influence of the *p*-cresol/2-propanol molar ratio on the toluene yield in order to limit hydrogenation of the aromatic rings. Furthermore, the efficiency of Ru/Nb_2O_5-SiO_2 catalyst on the CTH of other complex lignin model compounds (β-O-4 and α-O-4 linkages) was also studied [129].

4.2. CTH of Lignin

Lignin was efficiently converted into dimeric and monomeric compounds via the CTH approach by several research groups. The main obtained results are reported in Table 7.

Ford and co-workers demonstrated the occurrence of a single step process in the hydrogenolysis-depolymerization of the bio-oligomer organosolv lignin (obtained from sanded poplar sawdust) using a Cu-doped porous metal oxide (PMO) catalyst and supercritical MeOH (sc-MeOH) as a H-source/solvent at 300 °C for 24 h. A mixture of aromatics and monomeric substituted cyclohexyl derivates with a low oxygen content was formed together with a gas phase mainly composed of H_2 (79% mol), CH_4 (8%), CO (9%), and CO_2 (4%) [130].

Barta and co-workers reported the walnut lignin depolymerization using triflic acid as a catalyst in 1,4-dioxane at 140 °C for 4 h. The main reaction products were C2-aldehyde fragments [131].

Samec and co-workers reported also the transfer hydrogenolysis of Pine sawdust using the Pd/C catalyst and only an endogenous hydrogen source at 195 °C for 1 h converting its lignin content into aryl propene monomers. Formic acid generated from the organosolv process is the H-source used for the CTH reaction [132]. The same authors also propose the direct use of a hemicellulose fraction as an H-donor source to carry out CTH reactions [133]. Phenol, and saturated and unsaturated propylphenols were obtained as the main reaction products and their selectivity can be efficiently tuned by changing (i) the solvent biomass ratio and (ii) reaction temperatures/times.

Table 7. A literature overview on the CTH of lignin promoted by heterogeneous catalysts.

Entry	Lignin Type [1]	Catalyst	H-Source	Temp. [°C]	Time [h]	Conversion [%]	Main Products	Ref.
1	OL	Cu-MPO	MeOH	300	24	100	Cycloexyl derivates	[130]
2	DWL	Trifilic acid	1,4-dioxane	140	4	100	C2-aldehyde fragments	[131]
3	PS	Pd/C	Formic acid	195	1	100	Aryl propene monomers	[132]
4	OL	Pd/C	Hemicellulose	210	15	100	Phenols and propylphenols	[133]
5	OL	Pd/C	Carbohydrate fractions	200	2	100	4-ethylguaiacol	[134]
6	OL	Pd/C	Hemicellulose	160–220	3–6	100	Monophenolic products	[135]
7	OL	Ru/C	MeOH	250	3	90	4-n-propanolguaiacol and 4-n-propanolsyringol	[136]
8	OL	Ru/C	MeOH	250	3	85	para-propyl phenolics	[137]
9	OL	Ru/C	2-PO	300	1–3	100	4-ethyl phenol, 2-methoxy phenol and phenol	[137]
10	BVL	Ni/C	Aliphatic alcohols	200	6	50	4-propylguaiacol and 4-propylsyringol	[138]
11	OL	Al-SBA-15	Tetraline/FA	140	1/2	100	Mesitol and syrangaldehyde	[139]
12	OL	RANEY® Ni	Hemicellulose	160–220	3	100	Monocyclic products	[140]
13	OL	RANEY® Ni	2-PO/H₂O	160–220	18	100	Alkenes and arenes	[141]
14	KL	TiN-Ni and TiO₂-Ni	MeOH, EtOH,2-PO, THF	150	4.5 min	100	Guaiacol products	[142]
15	OL	Pd₃Ni₄/MIL-100(Fe)	H₂O	180	6	100	Phenol and guaiacol derivates	[143]
16	OL	FeB, NiB and FeNiB	EtOH	320	2	100	21 depolymerization products	[144]
17	KL	Fe on Rh/La₂O₃/CeO₂-ZrO₂	2-PO/H₂O	373	2	100	C₁₂₋₂₆ aliphatic, C₆₋₁₆ aromatic and C₇₋₁₀ hydrogenated cycles compounds	[145]
18	OL	Cu-Mg-Al oxides	EtOH	340	4	100	C₆₋₁₂ aromatics, C₃₋₈ alcohols, C₃₋₁₂ esters	[146]

[1] Abbreviation: OL: Organosolv Lignin; DVL: Dioxolv walnut Lignin; PS: Pine sawdust; BVL: Birch-wood Lignin; KL: Kraft Lignin.

In another recent contribution, the same research group presented a three-step process to convert lignin obtained from the *quercus suberin* to monomeric phenolic compounds and hydrocarbons by using the Pd/C catalyst in the CTH with carbohydrate fractions serving as hydrogen donors under mild alkaline conditions. The lignin was converted with a delignification (wt%) of 90%. [134].

In 2017, the same authors presented for the first time a method to fractionate the lignin in high yield to monophenolic products by a flow system in CTH conditions using the Pd/C catalyst and hemicellulose as an internal H-source and reducing agent at 160–220 °C for 3–6 h. Furthermore, the cellulose fraction was found as solid residue in 92 wt% [135].

Also, Sels and co-workers demonstrated that commercial Pd/C and Ru/C catalysts can be used in the CTH of lignin fractions (obtained from birch wood) using MeOH as an H-donor solvent [136]. They showed that Pd/C and Ru/C obtained different lignin products (4-n-propanolguaiacol and 4-n-propanolsyringol) at 250 °C for 3 h. The Ru/C catalyst favors the formation of para-propyl phenolics, while the Pd/C catalyst preferentially forms para-propanol phenol derivatives with a remarkable selectivity (91%) to 4-n-propanolguaiacol (PohG) and 4-n-propanolsyringol (PohS). Finally, they demonstred that the Pd/C catalyst is preferred when a high OH-content lignin oil is present.

Moreover, Kim and co-workers explored the potential efficiency of CTH using 2-propanol as a solvent/H-donor to valorize the lignin-rich residue obtained from an ionic liquid conversion process, using Ru/C (5% wt on C) as a catalyst. Monomeric and alkyl-substituted phenols (4-ethyl phenol, 2-methoxy phenol, and phenol) were the main reaction products in the liquid oil, suggesting that the lignin residue can be efficiently depolymerized under CTH conditions [137].

Song's research group proposed a successful method using the Ni/C catalyst in the presence of simple aliphatic alcohols as a hydrogen source. Under mild reaction conditions (200 °C, for 6 h, 1 MPa Ar), the birch-wood lignin was converted into 4-propylguaiacol (M7G) and 4-propylsyringol (M7S) with a selectivity of 97% and 54% yield referred to all monomers. They proposed that the lignin is first fragmented into smaller lignin species by alcoholysis reactions and then the Ni/C catalyst converts oligomers to monomeric phenols [138].

Toledano and Luque published a microwave hydrogenolytic method (400 W at 140 °C for 30 min) to depolymerise organosolv lignin, isolated from tree pruning, into simple phenolic compounds, including mesitol and syringaldehyde by mild hydrogen-free conditions, using a range of bifunctional catalysts based on metal supported nanoparticles (Ni, Ru, Pd, and Pt) on the mesoporous acidic aluminosilicate support (Al-SBA-15) with tetraline or formic acid as the H-donor/solvents. Among all catalysts, the Ni10%AlSBA gave the best degree of lignin depolymerization after only 30 min of reaction with the main products being bio-oil, bio-char, and residual lignin [139].

Raney Ni can favor the organosolv separation, promoting the upgrating of the liquor by means of an H-transfer process [140,141]. Similar to studies conducted by the Samec research group, the hemicellulose fraction acts as an H-donor substrate.

Conversely, Esposito et al. compared the effect of two different Ni-based systems (TiN-Ni and TiO$_2$-Ni) in the reductive depolymerization of Kraft lignin to substituted guaiacol products without molecular H$_2$ in a flow-reactor system (150 °C for ≈4.5 min) [142]. TiN-Ni shows a better catalytic performance than TiO$_2$-Ni as a consequence of the major dispersion of Ni in the TiN phase. At the same time, the TiN-Ni system presents a better stability than the well-known Raney Ni and Pd/C catalysts.

Cai and co-workers reported the excellent catalytic performance of the Pd$_1$Ni$_4$/MIL-100(Fe) catalyst in the self-hydrogenolysis of organosolv lignin, using water as solvent, at 180 °C for 6 h. A 17% monomer yield and a set of reaction products including substituted phenol and guaiacol derivatives were registered. The catalyst shows a highly porous structure, strong Lewis acid properties, and water stability. Furthermore, the catalytic system can be recycled up to five times [143].

Chmely and co-workers, examined the catalytic activity of three different nanomaterial, amorphous B-containing FeNi alloys (FeB, NiB and FeNiB) in the transfer hydrogenolysis reaction of organosolv lignin using supercrtical ethanol as an H-source/solvent. FeNiB shows the best reactivity (74% conversion) and selectivity (84%) among the three catalysts, affording 21 different depolymerization products [144].

Jin and co-workers described the depolymerization of the Kraft lignin using a 2-propanol/water mixture system as solvent/hydrogen source over Fe on $Rh/La_2O_3/CeO_2$-ZrO_2 catalyst at 373 °C for 2 h. The main products obtained were C_{12-26} aliphatic, C_{6-16} aromatic, and C_{7-10} hydrogenated cycles compounds [145].

Hensen et al. investigated the role of different Cu-Mg-Al mixed oxides as catalysts, prepared by varying the Cu content and the (Cu+Mg)/Al ratio, in the depolymerization of lignin in CTH conditions by using supercritical ethanol as a solvent and H-donor at 340 °C for 4 h. The optimum performance was given by the mixture containing 20 wt% Cu and having a (Cu+Mg)/Al ratio of 4. $Cu_{20}MgAl_4$ affords the highest monomers yield and the least amount of repolymerization products during the lignin conversion [146].

5. Conclusions and Perspectives

In this review we summarize recent reports concerning the conversion of cellulose, hemicellulose and lignin promoted by heterogeneous catalysts by using the transfer hydrogenolysis (CTH) approach. A lot of attention has been paid in the CTH of their relative derived molecules with polyols, furan derivatives, levulinic acid, and aromatic ethers being the most investigated in order to get a chemical insight in the cleavage of C–O and C–C bond cleavage by alternative H-sources. At present, two main H-donor molecules have mainly been used: simple primary-secondary alcohols (methanol, ethanol, and 2-propanol) and formic acid.

Surely, lignin depolymerization—through transfer hydrogenolysis processes—has been more investigated with respect to other components of lignocellulosic biomasses due to the possibility to produce more aromatic compounds with respect to analogous catalytic processes that traditionally use high-pressure hydrogen gas. At the same time, only a few examples are present in the CTH of cellulose in the literature, whereas hemicellulose has been proposed mainly as an H-source itself in the direct upgrading of woods via tandem organosolv-transfer hydrogenolysis reactions.

In summary, the CTH seems to have all the potential to be a valid alternative to the use of molecular hydrogen for the reductive upgrading of lignocellulosic biomasses; however, a better understanding of catalysts' active sites, together with the molecular level of both C–O and C–C bond-breaking mechanism, is still needed. In particular, although the development of several heterogeneous catalysts has made significant progress in widening the choice of the transfer hydrogenolysis chemistry allowing also an increase of reaction rates and yields that has for a long time hindered the CTH application in recent years, the role of active metal sites and acid-base sites also remains to be clarified. In order to define correct structure–activity relationships, a comprehensive knowledge of the nature of the active sites (metal, Brønsted or Lewis acid, or base site) is necessary, as well as their strength and density. The development of catalytic systems characterized by tailored site compositions and sharp structures, employed in investigative model reactions, will be useful in revealing the synergistic effects of the different types of active sites, also contributing in outlining the kinetic behavior and mechanistic features of the title reactions. A comprehensive understanding of the reaction pathway at the molecular level of the C–O and C–C bond-breaking mechanism, could be attained by using computational studies, isotopic labeling, and combined kinetic measurements. The use of innovative approaches, such as in situ and interfacial sensitive spectroscopic tools, capable of elucidating the active hydrogen species involved in the hydrogen transfer step, the reaction intermediates at solid/liquid interfaces, and probing how the presence of solvent affects the interaction between reactants and active sites, would be desirable. Furthermore, in order to make CTH processes competitive with high pressure hydrogen gas conventional procedures and also to have the possibility of getting a real outcome in biomass transformation, basic studies on the role of a solvent as an H-source, as well as the practical application of transfer hydrogenolysis processes in the one-pot direct conversion of real lignocellulosic biomasses, are still necessary tasks.

Author Contributions: All authors contributed equally to this review.

Funding: This publication was supported from Università Mediterranea di Reggio Calabria and PON ("ACQUASYSTEM" and "MEL" projects).

Acknowledgments: Authors thanks Pietropaolo and Signorino Galvagno for fruitful discussions.

Conflicts of Interest: The authors declare no conflict of interest.

References

1. Klass, D.L. *Biomass for Renewable Energy, Fuels, and Chemicals*; Academic Press: San Diego, CA, USA, 1998.
2. The White House-Washington. *National Bio-economy Blueprint*; The White House: Washington, DC, USA, 2012; pp. 1–43. Available online: https://obamawhitehouse.archives.gov/sites/default/files/microsites/ostp/national_bioeconomy_blueprint_april_2012.pdf (accessed on 30 June 2018).
3. European Commission. *Innovating for Sustainable Growth: A Bioeconomy for Europe*; European Commission: Brussels, Belgium, 2012; pp. 1–9. Available online: http://ec.europa.eu/research/bioeconomy/pdf/official-strategy_en.pdf (accessed on 30 June 2018).
4. European Commission. *A Roadmap for Moving to a Competitive Low Carbon Economy in 2050*; European Commission: Brussels, Belgium, 2011; Available online: http://eur-lex.europa.eu/legal-content/EN/ALL/?uri=CELEX:52011DC0112 (accessed on 30 December 2016).
5. Lee, D.-H. Bio-based economies in Asia: Economic analysis of development of bio-based industry in China, India, Japan, Korea, Malaysia and Taiwan. *Int. J. Hydrogen Energy* **2016**, *41*, 4333–4346. [CrossRef]
6. Dey, S. Asian bioeconomy and biobusiness: Current scenario and future prospects. *New Biotechnol.* **2014**, *31*, S34. [CrossRef]
7. Ignaciuk, A.; Vöhringer, F.; Ruijs, A.; Van Ierland, E.C. Competition between biomass and food production in the presence of energy policies: A partial equilibrium analysis. *Energy Policy* **2006**, *34*, 1127–1138. [CrossRef]
8. Somerville, C.; Youngs, H.; Taylor, C.; Davis, S.C.; Long, S.P. Feedstocks for lignocellulosic biofuels. *Science* **2010**, *329*, 790–792. [CrossRef] [PubMed]
9. Tuck, C.O.; Pérez, E.; Horváth, I.T.; Sheldon, R.A.; Poliakoff, M. Valorization of biomass: Deriving more value from waste. *Science* **2012**, *337*, 695–699. [CrossRef] [PubMed]
10. Yan, K.; Yang, Y.; Chai, J.; Lu, Y. Catalytic reactions of gamma-valerolactone: A platform to fuels and value-added chemicals. *Appl. Catal. B Environ.* **2015**, *179*, 292–304. [CrossRef]
11. Hu, L.; Lin, L.; Wu, Z.; Zhou, S.; Liu, S. Chemocatalytic hydrolysis of cellulose into glucose over solid acid catalysts. *Appl. Catal. B Environ.* **2015**, *174–175*, 225–243. [CrossRef]
12. Negahdar, L.; Delidovich, I.; Palkovits, R. Aqueous-phase hydrolysis of cellulose and hemicelluloses over molecular acidic catalysts: Insights into the kinetics and reaction mechanism. *Appl. Catal. B Environ.* **2016**, *184*, 285–298. [CrossRef]
13. Sheldon, R.A. Green and sustainable manufacture of chemicals from biomass: State of the art. *Green Chem.* **2014**, *16*, 950–963. [CrossRef]
14. Zhang, Z.; Song, J.; Han, B. Catalytic transformation of lignocellulose into chemicals and fuel products in ionic liquids. *Chem. Rev.* **2016**, *117*, 6834–6880. [CrossRef] [PubMed]
15. Li, C.; Zhao, X.; Wang, A.; Huber, G.H.; Zhang, T. Catalytic transformation of lignin for the production of chemicals and fuels. *Chem. Rev.* **2015**, *115*, 11559–11624. [CrossRef] [PubMed]
16. Lange, J.P. Lignocellulose conversion: An introduction to chemistry, process and economics. *Biofuels Bioprod. Biorefin.* **2007**, *1*, 39–48. [CrossRef]
17. Stöcker, M. Biofuels and biomass-to-liquid fuels in the biorefinery: Catalytic conversion of lignocellulosic biomass using porous materials. *Angew. Chem. Int. Ed.* **2008**, *47*, 9200–9211. [CrossRef] [PubMed]
18. Espro, C.; Gumina, B.; Paone, E.; Mauriello, F. Upgrading Lignocellulosic Biomasses: Hydrogenolysis of Platform Derived Molecules Promoted by Heterogeneous Pd-Fe Catalysts. *Catalysts* **2017**, *7*, 78. [CrossRef]
19. Sun, Z.; Fridrich, B.; De Santi, A.; Elangovan, S.; Barta, K. Bright Side of Lignin Depolymerization: Toward New Platform Chemicals. *Chem. Rev.* **2018**, *118*, 614–678. [CrossRef] [PubMed]
20. Schutyser, W.; Renders, T.; Van den Bosch, S.; Koelewijn, S.-F.; Beckham, G.T.; Sels, B.F. Chemicals from lignin: An interplay of lignocellulose fractionation, depolymerisation, and upgrading. *Chem. Soc. Rev.* **2018**, *47*, 852–908. [CrossRef] [PubMed]

21. Global Bio-Based Chemical Market Forecast 2018–2026. Available online: https://www.reportlinker.com/p05001382/Global-Bio-Based-Chemicals-Market-Forecast.html (accessed on 30 June 2018).

22. Dornburg, V.; Hermann, B.G.; Patel, M.K. Scenario Projections for Future Market Potentials of Biobased Bulk Chemicals. *Environ. Sci. Technol.* **2008**, *42*, 2261–2267. [CrossRef] [PubMed]

23. Chen, H. Chemical composition and structure of natural lignocellulose. In *Biotechnology of Lignocellulose: Theory and Practice*; Springer Science + Business Media B.V.: Dordrecht, The Netherlands, 2014; pp. 25–71.

24. Klemm, D.; Heublein, B.; Fink, H.-P.; Bohn, A. Cellulose: Fascinating biopolymer and sustainable raw material. *Angew. Chem. Int. Ed.* **2005**, *44*, 3358–3393. [CrossRef] [PubMed]

25. Kobayashi, H.; Fukuoka, A. Synthesis and utilisation of sugar compounds derived from lignocellulosic biomass. *Green Chem.* **2013**, *15*, 1740–1763. [CrossRef]

26. Isikgor, F.H.; Becer, C.R. Lignocellulosic Biomass: A sustainable platform for production of bio-based chemicals and polymers. *Polym. Chem.* **2015**, *6*, 4497–4559. [CrossRef]

27. Besson, M.; Gallezot, P.; Pinel, C. Conversion of biomass into chemicals over metal catalysts. *Chem. Rev.* **2014**, *114*, 1827–1870. [CrossRef] [PubMed]

28. Corma, A.; Iborra, S.; Velty, A. Chemical routes for the transformation of biomass into chemicals. *Chem. Rev.* **2007**, *107*, 2411–2502. [CrossRef] [PubMed]

29. Binder, J.B.; Raines, R.T. Simple Chemical Transformation of Lignocellulosic Biomass into Furans for Fuels and Chemicals. *J. Am. Chem. Soc.* **2009**, *131*, 1979–1985. [CrossRef] [PubMed]

30. Xu, C.; Arancon, R.A.D.; Labidi, J.; Luque, R. Lignin depolymerisation strategies: Towards valuable chemicals and fuels. *Chem. Soc. Rev.* **2014**, *43*, 7485–7500. [CrossRef] [PubMed]

31. Deuss, P.J.; Barta, K. From models to lignin: Transition metal catalysis for selective bond cleavage reactions. *Coord. Chem. Rev.* **2016**, *306*, 510–532. [CrossRef]

32. Zakzeski, J.; Bruijnincx, P.C.A.; Jongerius, A.L.; Weckhuysen, B.M. The Catalytic Valorization of Lignin for the Production of Renewable Chemicals. *Chem. Rev.* **2010**, *110*, 3552–3599. [CrossRef] [PubMed]

33. Galkin, M.V.; Samec, J.S.M. Lignin Valorization through Catalytic Lignocellulose Fractionation: A Fundamental Platform for the Future Biorefinery. *ChemSusChem* **2016**, *9*, 1544–1558. [CrossRef] [PubMed]

34. Bridgwater, A.V.; Meierb, D.; Radlein, D. An overview of fast pyrolysis of biomass. *Org. Geochem.* **1999**, *30*, 1479–1493. [CrossRef]

35. De, S.; Saha, B.; Luque, R. Hydrodeoxygenation processes: Advances on catalytic transformations of biomass-derived platform chemicals into hydrocarbon fuels. *Bioresour. Technol.* **2015**, *178*, 108–118. [CrossRef] [PubMed]

36. Ruppert, A.M.; Weinberg, K.; Palkovits, R. Hydrogenolysis goes bio: From carbohydrates and sugar alcohols to platform chemicals. *Angew. Chem. Int. Ed.* **2012**, *51*, 2564–2601. [CrossRef] [PubMed]

37. Li, N.; Wang, W.; Zheng, M.; Zhang, T. General Reaction Mechanisms in Hydrogenation and Hydrogenolysis for Biorefining. In *Catalytic Hydrogenation for Biomass Valorization*; Rinaldi, R., Ed.; Royal Society of Chemistry: Cambridge, UK, 2015; pp. 22–50.

38. Wang, D.; Astruc, D. The Golden Age of Transfer Hydrogenation. *Chem. Rev.* **2015**, *115*, 6621–6686. [CrossRef] [PubMed]

39. Gilkey, M.J.; Xu, B. Heterogeneous Catalytic Transfer Hydrogenation as an Effective Pathway in Biomass Upgrading. *ACS Catal.* **2016**, *6*, 1420–1436. [CrossRef]

40. Muzart, J. Pd-Catalyzed Hydrogen-Transfer Reactions from Alcohols to C=C, C=O and C=N Bonds. *Eur. J. Org. Chem.* **2015**, *2015*, 5693–5707. [CrossRef]

41. Barta, K.; Ford, P.C. Catalytic Conversion of Nonfood Woody Biomass Solids to Organic Liquids. *Acc. Chem. Res.* **2014**, *47*, 1503–1512. [CrossRef] [PubMed]

42. Mellmann, D.; Sponholz, P.; Junge, H.; Beller, M. Formic acid as a hydrogen storage material–development of homogeneous catalysts for selective hydrogen release. *Chem. Soc. Rev.* **2016**, *45*, 3954–3988. [CrossRef] [PubMed]

43. Loges, B.; Boddien, A.; Gärtner, F.; Junge, H.; Beller, M. Catalytic Generation of Hydrogen from Formic acid and its Derivatives: Useful Hydrogen Storage Materials. *Top. Catal.* **2010**, *53*, 902–914. [CrossRef]

44. Grasemann, M.; Laurenczy, G.; Hirose, T.; Raspail, P.; Liu, S.L.; Wu, Y.P.; Guo, Q.-X.; Ludwig, R.; Beller, M. Formic acid as a hydrogen source–recent developments and future trends. *Energy Environ. Sci.* **2012**, *5*, 8171–8181. [CrossRef]

45. Musolino, M.G.; Scarpino, L.A.; Mauriello, F.; Pietropaolo, R. Selective transfer hydrogenolysis of glycerol promoted by palladium catalysts in absence of hydrogen. *Green Chem.* **2009**, *11*, 1511–1513. [CrossRef]
46. Mauriello, F.; Ariga, H.; Musolino, M.G.; Pietropaolo, R.; Takakusagi, S.; Asakura, K. Exploring the catalytic properties of supported palladium catalysts in the transfer hydrogenolysis of glycerol. *Appl. Catal. B Environ.* **2015**, *166–167*, 121–131. [CrossRef]
47. Gandarias, I.; Arias, P.L.; Requies, J.; El Doukkali, M.; Güemez, M.B. Liquid-phase glycerol hydrogenolysis to 1,2-propanediol under nitrogen pressure using 2-propanol as hydrogen source. *J. Catal.* **2011**, *282*, 237–247. [CrossRef]
48. Gandarias, I.; Arias, P.L.; Fernández, S.G.; Requies, J.; El Doukkali, M.; Güemez, M.B. Hydrogenolysis through catalytic transfer hydrogenation: Glycerol conversion to 1,2-propanediol. *Catal. Today* **2012**, *195*, 22–31. [CrossRef]
49. Gandarias, I.; Requies, J.; Arias, P.L.; Armbruster, U.; Martin, A. Liquid-phase glycerol hydrogenolysis by formic acid over Ni–Cu/Al$_2$O$_3$ catalysts. *J. Catal.* **2012**, *290*, 79–89. [CrossRef]
50. Mane, R.B.; Rode, C.V. Continuous Dehydration and Hydrogenolysis of Glycerol over Non-Chromium Copper Catalyst: Laboratory-Scale Process Studies. *Org. Process Res. Dev.* **2012**, *16*, 1043–1052. [CrossRef]
51. Yuan, J.; Li, S.; Yu, L.; Liu, Y.; Cao, Y. Efficient catalytic hydrogenolysis of glycerol using formic acid as hydrogen source. *Chin. J. Catal.* **2013**, *34*, 2066–2074. [CrossRef]
52. Xia, S.; Yuan, Z.; Wang, L.; Chen, P.; Hou, Z. Hydrogenolysis of glycerol on bimetallic Pd-Cu/solid-base catalysts prepared via layered double hydroxides precursors. *Appl. Catal. A Gen.* **2011**, *403*, 173–182. [CrossRef]
53. Bienholz, A.; Hofmann, H.; Claus, P. Selective hydrogenolysis of glycerol over copper catalysts both in liquid and vapour phase: Correlation between the copper surface area and the catalyst's activity. *Appl. Catal. A Gen.* **2011**, *391*, 153–157. [CrossRef]
54. Xia, S.; Yuan, Z.; Wang, L.; Chen, P.; Hou, Z. Catalytic production of 1,2-propanediol from glycerol in bio-ethanol solvent. *Bioresour. Technol.* **2012**, *104*, 814–817. [CrossRef] [PubMed]
55. Musolino, M.G.; Scarpino, L.A.; Mauriello, F.; Pietropaolo, R. Glycerol hydrogenolysis promoted by supported palladium catalysts. *ChemSusChem* **2011**, *4*, 1143–1150. [CrossRef] [PubMed]
56. Gong, L.; Lü, Y.; Ding, Y.; Lin, R.; Li, J.; Dong, W.; Wang, T.; Chen, W. Solvent Effect on Selective Dehydroxylation of Glycerol to 1,3-Propanediol over a Pt/WO$_3$/ZrO$_2$ Catalyst. *Chin. J. Catal.* **2009**, *30*, 1189–1191. [CrossRef]
57. Liu, Y.; Tüysü, H.; Jia, C.-J.; Schwickardi, M.; Rinaldi, R.; Lu, A.-H.; Schmidt, W.; Schüt, F. From glycerol to allyl alcohol: Iron oxide catalyzed dehydration and consecutive hydrogen transfer. *Chem. Commun.* **2010**, *46*, 1238–1240. [CrossRef] [PubMed]
58. Konaka, A.; Tago, T.; Yoshikawa, T.; Nakamura, A.; Masuda, T. Conversion of glycerol into allyl alcohol over potassium-supported zirconia–iron oxide catalyst. *Appl. Catal. B Environ.* **2014**, *146*, 267–273. [CrossRef]
59. Arceo, E.; Marsden, P.; Bergman, R.G.; Ellman, J.A. An efficient didehydroxylation method for the biomass-derived polyols glycerol and erythritol. Mechanistic studies of a formic acid-mediated deoxygenation. *Chem. Commun.* **2009**, *23*, 3357–3359. [CrossRef] [PubMed]
60. Dethlefsen, J.R.; Lupp, D.; Teshome, A.; Nielsen, L.B.; Fristrup, P. Molybdenum-catalyzed conversion of diols and biomass-derived polyols to alkenes using isopropyl alcohol as reductant and solvent. *ACS Catal.* **2015**, *5*, 3638–3647. [CrossRef]
61. Deng, L.; Li, J.; Lai, D.-M.; Fu, Y.; Guo, Q.-X. Catalytic conversion of biomass-derived carbohydrates into γ-valerolactone without using an external H$_2$ supply. *Angew. Chem. Int. Ed.* **2009**, *48*, 6529–6532. [CrossRef] [PubMed]
62. Heeres, H.; Handana, R.; Chunai, D.; Rasrendra, C.B.; Girisuta, B.; Heeres, H.J. Combined dehydration/(transfer)-hydrogenation of C6-sugars (D-glucose and D-fructose) to γ-valerolactone using ruthenium catalysts. *Green Chem.* **2009**, *11*, 1247–1255. [CrossRef]
63. Du, X.-L.; He, L.; Zhao, S.; Liu, Y.-M.; Cao, Y.; He, H.-Y.; Fan, K.-N. Hydrogen-independent reductive transformation of carbohydrate biomass into γ-valerolactone and pyrrolidone derivatives with supported gold catalysts. *Angew. Chem.* **2011**, *123*, 7961–7965. [CrossRef]
64. Scholz, D.; Aellig, C.; Mondelli, C.; Pèrez-Ramìrez, J. Continuous transfer hydrogenation of sugars to alditols with bioderived donors over Cu–Ni–Al catalysts. *ChemCatChem* **2015**, *7*, 1551–1558. [CrossRef]

65. Van Hengstum, A.J.; Kieboom, A.P.G.; van Bekkum, H. Catalytic transfer hydrogenation of glucose-fructose syrups in alkaline solution. *Starch* **1984**, *36*, 317–320. [CrossRef]

66. Kobayashi, H.; Matsuhashi, H.; Komanoya, T.; Hara, K.; Fukuoka, A. Transfer hydrogenation of cellulose to sugar alcohols over supported ruthenium catalysts. *Chem. Commun.* **2011**, *47*, 2366–2368. [CrossRef] [PubMed]

67. Shrotri, A.; Kobayashi, H.; Tanksale, A.; Fukuoka, A.; Beltramini, J. Transfer Hydrogenation of Cellulose-based Oligomers over Carbon-supported Ruthenium Catalyst in a Fixed-bed Reactor. *ChemCatChem* **2014**, *6*, 1349–1356. [CrossRef]

68. Dalvand, K.; Rubin, J.; Gunukula, S.; Clayton Wheeler, M.; Hunt, G. Economics of biofuels: Market potential of furfural and its derivatives. *Biomass Bioenergy* **2018**, *115*, 56–63. [CrossRef]

69. Mariscal, R.; Maireles-Torres, P.; Ojeda, M.; Sádaba, I.; López Granados, M. Furfural: A renewable and versatile platform molecule for the synthesis of chemicals and fuels. *Energy Environ. Sci.* **2016**, *9*, 1144–1189. [CrossRef]

70. Li, X.; Jia, P.; Wang, T. Furfural: A Promising Platform Compound for Sustainable Production of C4 and C5 Chemicals. *ACS Catal.* **2016**, *6*, 7621–7640. [CrossRef]

71. Li, J.; Liu, J.; Zhou, H.; Fu, Y. Catalytic Transfer Hydrogenation of Furfural to Furfuryl Alcohol over Nitrogen-Doped Carbon-Supported Iron Catalysts. *ChemSusChem* **2016**, *9*, 1339–1347. [CrossRef] [PubMed]

72. Tang, X.; Wei, J.; Ding, N.; Sun, Y.; Zeng, X.; Hu, L.; Liu, S.; Lei, T.; Lin, L. Chemoselective hydrogenation of biomass derived 5-hydroxymethylfurfural to diols: Key intermediates for sustainable chemicals, materials and fuels. *Renew. Sustain. Energy Rev.* **2017**, *77*, 287–296. [CrossRef]

73. Panagiotopoulou, P.; Martin, N.; Vlachos, D.G. Effect of hydrogen donor on liquid phase catalytic transfer hydrogenation of furfural over a Ru/RuO2/C catalyst. *J. Mol. Catal. A Chem.* **2014**, *392*, 223–228. [CrossRef]

74. Gilkey, M.J.; Panagiotopoulou, P.; Mironenko, A.V.; Jenness, G.R.; Vlachos, D.G.; Xu, B. Mechanistic Insights into Metal Lewis Acid-Mediated Catalytic Transfer Hydrogenation of Furfural to 2-Methylfuran. *ACS Catal.* **2015**, *5*, 3988–3994. [CrossRef]

75. Panagiotopoulou, P.; Vlachos, D.G. Liquid phase catalytic transfer hydrogenation of furfural over a Ru/C catalyst. *Appl. Catal. A Gen.* **2014**, *480*, 17–24. [CrossRef]

76. Wang, B.; Li, C.; He, B.; Qi, J.; Liang, C. Highly stable and selective Ru/NiFe2O4 catalysts for transfer hydrogenation of biomass-derived furfural to 2-methylfuran. *J. Energy Chem.* **2017**, *26*, 799–807. [CrossRef]

77. Zhang, Z.; Pei, Z.; Chen, H.; Chen, K.; Hou, Z.; Lu, X.; Ouyang, P.; Fu, J. Catalytic in-Situ Hydrogenation of Furfural over Bimetallic Cu-Ni Alloy Catalysts in Isopropanol. *Ind. Eng. Chem. Res.* **2018**, *57*, 4225–4230. [CrossRef]

78. Gong, W.; Chen, C.; Fan, R.; Zhang, H.; Wang, G.; Zhao, H. Transfer-hydrogenation of furfural and levulinic acid over supported copper catalyst. *Fuel* **2018**, *231*, 165–171. [CrossRef]

79. Chang, X.; Liu, A.-F.; Cai, B.; Luo, J.-Y.; Pan, H.; Huang, Y.-B. Catalytic Transfer Hydrogenation of Furfural to 2-Methylfuran and 2-Methyltetrahydrofuran over Bimetallic Copper-Palladium Catalysts. *ChemSusChem* **2016**, *9*, 3330–3337. [CrossRef] [PubMed]

80. Zhang, J.; Chen, J. Selective Transfer Hydrogenation of Biomass-Based Furfural and 5-Hydroxymethylfurfural over Hydrotalcite-Derived Copper Catalysts Using Methanol as a Hydrogen Donor. *ACS Sustain. Chem. Eng.* **2017**, *5*, 5982–5993. [CrossRef]

81. Scholz, D.; Aellig, C.; Hermans, I. Catalytic Transfer Hydrogenation/Hydrogenolysis for Reductive Upgrading of Furfural and 5-(Hydroxymethyl)furfural. *ChemSusChem* **2014**, *7*, 268–275. [CrossRef] [PubMed]

82. Yang, P.; Xia, Q.; Liu, X.; Wang, Y. Catalytic transfer hydrogenation/hydrogenolysis of 5-hydroxymethylfurfural to 2,5-dimethylfuran over Ni-Co/C catalyst. *Fuel* **2017**, *187*, 159–166. [CrossRef]

83. Qi, L.; Mui, Y.F.; Lo, S.W.; Lui, M.Y.; Akien, G.R.; Horváth, I.T. Catalytic conversion of fructose, glucose, and sucrose to 5-(hydroxymethyl)furfural and levulinic and formic acids in γ-valerolactone as a green solvent. *ACS Catal.* **2014**, *4*, 1470–1477. [CrossRef]

84. Jae, J.; Zheng, W.; Lobo, R.F.; Vlachos, D.G. Production of dimethylfuran from hydroxymethylfurfural through catalytic transfer hydrogenation with ruthenium supported on carbon. *ChemSusChem* **2013**, *6*, 1158–1162. [CrossRef] [PubMed]

85. Wang, T.; Zhang, J.; Xie, W.; Tang, Y.; Guo, D.; Ni, Y. Catalytic Transfer Hydrogenation of Biobased HMF to 2,5-Bis-(Hydroxymethyl)Furan over Ru/Co3O4. *Catalysts* **2017**, *7*, 92. [CrossRef]

86. Aellig, C.; Jenny, F.; Scholz, D.; Wolf, P.; Giovinazzo, I.; Kollhoff, F.; Hermans, I. Combined 1,4-butanediol lactonization and transfer hydrogenation/hydrogenolysis of furfural-derivatives under continuous flow conditions. *Catal. Sci. Technol.* **2014**, *4*, 2326–2331. [CrossRef]

87. Hansen, T.S.; Barta, K.; Anastas, P.T.; Ford, P.C.; Riisager, A. One-pot reduction of 5-hydroxymethylfurfural via hydrogen transfer from supercritical methanol. *Green Chem.* **2012**, *14*, 2457–2461. [CrossRef]

88. Jae, J.; Mahmoud, E.; Lobo, R.F.; Vlachos, D.G. Cascade of liquid-phase catalytic transfer hydrogenation and etherification of 5-hydroxymethylfurfural to potential biodiesel components over Lewis acid zeolites. *ChemCatChem* **2014**, *6*, 508–513. [CrossRef]

89. Hao, W.; Li, W.; Tang, X.; Zeng, X.; Sun, Y.; Liu, S.; Lin, L. Catalytic transfer hydrogenation of biomass-derived 5-hydroxymethyl furfural to the building block 2,5-bishydroxymethyl furan. *Green Chem.* **2016**, *18*, 1080–1088. [CrossRef]

90. Hu, L.; Yang, M.; Xu, N.; Xu, J.; Zhou, S.; Chu, X.; Zhao, Y. Selective transformation of biomass-derived 5-hydroxymethylfurfural into 2,5-dihydroxymethylfuran via catalytic transfer hydrogenation over magnetic zirconium hydroxides. *Korean J. Chem. Eng.* **2018**, *35*, 99–109. [CrossRef]

91. Pasini, T.; Lolli, A.; Albonetti, S.; Cavani, F.; Mella, M. Methanol as a clean and efficient H-transfer reactant for carbonyl reduction: Scope, limitations, and reaction mechanism. *J. Catal.* **2014**, *317*, 206–219. [CrossRef]

92. Thananatthanachon, T.; Rauchfuss, T.B. Efficient production of the liquid fuel 2,5-dimethylfuran from fructose using formic acid as a reagent. *Angew. Chem. Int. Ed.* **2010**, *49*, 6616–6618. [CrossRef] [PubMed]

93. Thananatthanachon, T.; Rauchfuss, T.B. Efficient route to hydroxymethylfurans from sugars via transfer hydrogenation. *ChemSusChem* **2010**, *3*, 1139–1141. [CrossRef] [PubMed]

94. Tuteja, J.; Choudhary, H.; Nishimura, S.; Ebitani, K. Direct Synthesis of 1,6-Hexanediol from HMF over a Heterogeneous Pd/ZrP Catalyst using Formic Acid as Hydrogen Source. *ChemSusChem* **2014**, *7*, 96–100. [CrossRef] [PubMed]

95. Gao, Z.; Li, C.; Fan, G.; Yang, L.; Li, F. Nitrogen-doped carbon-decorated copper catalyst for highly efficient transfer hydrogenolysis of 5-hydroxymethylfurfural to convertibly produce 2,5-dimethylfuran or 2,5-dimethyltetrahydrofuran. *Appl. Catal. B Environ.* **2018**, *226*, 523–533. [CrossRef]

96. Bozell, J.J.; Moens, L.; Elliott, D.; Wang, Y.; Neuenscwander, G.; Fitzpatrick, S.; Bilski, R.; Jarnefeld, J. Production of levulinic acid and use as a platform chemical for derived products. *Resour. Conserv. Recycl.* **2000**, *28*, 227–239. [CrossRef]

97. Antonetti, C.; Licursi, D.; Fulignati, S.; Valentini, G.; Raspolli Galletti, A. New Frontiers in the Catalytic Synthesis of Levulinic Acid: From Sugars to Raw and Waste Biomass as Starting Feedstock. *Catalysts* **2016**, *6*, 196. [CrossRef]

98. Hengne, A.M.; Kadu, B.S.; Biradar, N.S.; Chikate, R.C.; Rode, C.V. Transfer hydrogenation of biomass-derived levulinic acid to γ-valerolactone over supported Ni catalysts. *RSC Adv.* **2016**, *6*, 59753–59761. [CrossRef]

99. Yang, Z.; Huang, Y.B.; Guo, Q.X.; Fu, Y. RANEY® Ni catalyzed transfer hydrogenation of levulinate esters to γ-valerolactone at room temperature. *Chem. Commun.* **2013**, *49*, 5328–5330. [CrossRef] [PubMed]

100. Song, J.; Wu, L.; Zhou, B.; Zhou, H.; Fan, H.; Yang, Y.; Meng, Q.; Han, B. A new porous Zr-containing catalyst with a phenate group: An efficient catalyst for the catalytic transfer hydrogenation of ethyl levulinate to γ-valerolactone. *Green Chem.* **2015**, *17*, 1626–1632. [CrossRef]

101. Tang, X.; Hu, L.; Sun, Y.; Zhao, G.; Hao, W.; Lin, L. Conversion of biomass-derived ethyl levulinate into γ-valerolactone via hydrogen transfer from supercritical ethanol over a ZrO$_2$ catalyst. *RSC Adv.* **2013**, *3*, 10277–10284. [CrossRef]

102. Tang, X.; Chen, H.; Hu, L.; Hao, W.; Sun, Y.; Zeng, X.; Lin, L.; Liu, S. Conversion of biomass to γ-valerolactone by catalytic transfer hydrogenation of ethyl levulinate over metal hydroxides. *Appl. Catal. B Environ.* **2014**, *147*, 827–834. [CrossRef]

103. Li, F.; France, L.J.; Cai, Z.; Li, Y.; Liu, S.; Lou, H.; Long, J.; Li, X. Catalytic transfer hydrogenation of butyl levulinate to Γ-valerolactone over zirconium phosphates with adjustable Lewis and Brønsted acid sites. *Appl. Catal. B Environ.* **2017**, *214*, 67–77. [CrossRef]

104. Kuwahara, Y.; Kaburagi, W.; Osada, Y.; Fujitani, T.; Yamashita, H. Catalytic transfer hydrogenation of biomass-derived levulinic acid and its esters to γ-valerolactone over ZrO$_2$ catalyst supported on SBA-15 silica. *Catal. Today* **2017**, *281*, 418–428. [CrossRef]

105. Wang, J.; Jaenicke, S.; Chuah, G.-K. Zirconium–Beta zeolite as a robust catalyst for the transformation of levulinic acid to γ-valerolactone via Meerwein–Ponndorf–Verley reduction. *RSC Adv.* **2014**, *4*, 13481–13489. [CrossRef]

106. Chia, M.; Dumesic, J.A. Liquid-phase catalytic transfer hydrogenation and cyclization of levulinic acid and its esters to γ-valerolactone over metal oxide catalysts. *Chem. Commun.* **2011**, *47*, 12233–12235. [CrossRef] [PubMed]

107. Bui, L.; Luo, H.; Gunther, W.R.; Román-Leshkov, Y. Domino Reaction Catalyzed by Zeolites with Brønsted and Lewis Acid Sites for the Production of γ-Valerolactone from Furfural. *Angew. Chem. Int. Ed.* **2013**, *52*, 8022–8025. [CrossRef] [PubMed]

108. Valekar, A.H.; Cho, K.H.; Chitale, S.K.; Hong, D.Y.; Cha, G.Y.; Lee, U.H.; Hwang, D.W.; Serre, C.; Chang, J.S.; Hwang, Y.K. Catalytic transfer hydrogenation of ethyl levulinate to γ-valerolactone over zirconium-based metal-organic frameworks. *Green Chem.* **2016**, *18*, 4542–4552. [CrossRef]

109. Tang, X.; Zeng, X.; Li, Z.; Hu, L.; Sun, Y.; Liu, S.; Lei, T.; Lin, L. Production of γ-valerolactone from lignocellulosic biomass for sustainable fuels and chemicals supply. *Renew. Sustain. Energy Rev.* **2014**, *40*, 608–620. [CrossRef]

110. Ortiz-Cervantes, C.; García, J.J. Hydrogenation of levulinic acid to γ-valerolactone using ruthenium nanoparticles. *Inorg. Chim. Acta* **2013**, *397*, 124–128. [CrossRef]

111. Deng, L.; Zhao, Y.; Li, J.; Fu, Y.; Liao, B.; Guo, Q.-X. Conversion of Levulinic Acid and Formic Acid into γ-Valerolactone over Heterogeneous Catalysts. *ChemSusChem* **2010**, *3*, 1172–1175. [CrossRef] [PubMed]

112. Son, P.A.; Nishimura, S.; Ebitani, K. Production of γ-valerolactone from biomass-derived compounds using formic acid as a hydrogen source over supported metal catalysts in water solvent. *RSC Adv.* **2014**, *4*, 10525–10530. [CrossRef]

113. Lomate, S.; Sultana, A.; Fujitani, T. Vapor Phase Catalytic Transfer Hydrogenation (CTH) of Levulinic Acid to γ-Valerolactone over Copper Supported Catalysts Using Formic Acid as Hydrogen Source. *Catal. Lett.* **2018**, *148*, 348–358. [CrossRef]

114. Yuan, J.; Li, S.-S.; Yu, L.; Liu, Y.-M.; Cao, Y.; He, H.-Y.; Fan, K.-N. Copper-based catalysts for the efficient conversion of carbohydrate biomass into γ-valerolactone in the absence of externally added hydrogen. *Energy Environ. Sci.* **2013**, *6*, 3308–3313. [CrossRef]

115. Hengne, A.M.; Malawadkar, A.V.; Biradar, N.S.; Rode, C.V. Surface synergism of an Ag–Ni/ZrO$_2$ nanocomposite for the catalytic transfer hydrogenation of bio-derived platform molecules. *RSC Adv.* **2014**, *4*, 9730–9736. [CrossRef]

116. Mascala, S.M.; Matson, T.D.; Johnson, C.L.; Lewis, R.S.; Iretskii, A.V.; Ford, P.C. Hydrogen Transfer from Supercritical Methanol over a Solid Base Catalyst: A Model for Lignin Depolymerization. *ChemSusChem* **2009**, *2*, 215–217. [CrossRef] [PubMed]

117. Besse, X.; Schuurman, Y.; Guilhame, N. Reactivity of lignin model compounds through hydrogen transfer catalysis in ethanol/water mixtures. *Appl. Catal. B* **2017**, *209*, 265–272. [CrossRef]

118. Wu, H.; Song, J.; Xie, C.; Wu, C.; Chen, C.; Han, B. Efficient and Mild Transfer Hydrogenolytic Cleavage of Aromatic Ether Bonds in Lignin-Derived Compounds over Ru/C. *ACS Sustain. Chem. Eng.* **2018**, *6*, 2872–2877. [CrossRef]

119. Galkin, M.V.; Sawadjoon, S.; Rohde, V.; Dawange, M.; Samec, J.S.M. Mild Heterogeneous Palladium-Catalyzed Cleavage of β-O-4'-Ether Linkages of Lignin Model Compounds and Native Lignin in Air. *ChemCatChem* **2014**, *6*, 179–184. [CrossRef]

120. Sawadjoon, S.; Lundstedt, A.; Samec, J.S.M. Pd-Catalyzed Transfer Hydrogenolysis of Primary, Secondary, and Tertiary Benzylic Alcohols by Formic Acid: A Mechanistic Study. *ACS Catal.* **2013**, *3*, 635–642. [CrossRef]

121. Zhang, D.; Ye, F.; Xue, T.; Guan, Y.; Wang, Y.M. Transfer hydrogenation of phenol on supported Pd catalysts using formic acid as an alternative hydrogen source. *Catal. Today* **2014**, *234*, 133–138. [CrossRef]

122. Wang, X.; Rinaldi, R. Exploiting H-transfer reactions with RANEY® Ni for upgrade of phenolic and aromatic biorefinery feeds under unusual, low-severity conditions. *Energy Environ. Sci.* **2012**, *5*, 8244–8260. [CrossRef]

123. Wang, X.; Rinaldi, R. A Route for Lignin and Bio-Oil Conversion: Dehydroxylation of Phenols into Arenes by Catalytic Tandem Reactions. *Angew. Chem. Int. Ed.* **2013**, *52*, 11499–11503. [CrossRef] [PubMed]

124. Kennema, M.; de Castro, I.B.D.; Meemken, F.; Rinaldi, R. Liquid-Phase H-Transfer from 2-Propanol to Phenol on Raney Ni: Surface Processes and Inhibition. *ACS Catal.* **2017**, *7*, 2437–2445. [CrossRef]

125. Mauriello, F.; Paone, E.; Pietropaolo, R.; Balu, A.M.; Luque, R. Catalytic transfer hydrogenolysis of lignin derived aromatic ethers promoted by bimetallic Pd/Ni systems. *ACS Sustain. Chem. Eng.* **2018**, *6*, 9269–9276. [CrossRef]

126. Cozzula, D.; Vinci, A.; Mauriello, F.; Pietropaolo, R.; Müller, T.E. Directing the Cleavage of Ester C–O Bonds by Controlling the Hydrogen Availability on the Surface of Coprecipitated Pd/Fe$_3$O$_4$. *ChemCatChem* **2016**, *8*, 1515–1522. [CrossRef]

127. Paone, E.; Espro, C.; Pietropaolo, R.; Mauriello, F. Selective arene production from transfer hydrogenolysis of benzyl phenyl ether promoted by a co-precipitated Pd/Fe$_3$O$_4$ catalyst. *Catal. Sci. Technol.* **2016**, *6*, 7937–7941. [CrossRef]

128. Kim, M.; Ha, J.-M.; Lee, K.-Y.; Jae, J. Catalytic transfer hydrogenation/hydrogenolysis of guaiacol to cyclohexane over bimetallic RuRe/C catalysts. *Catal. Commun.* **2016**, *86*, 113–118. [CrossRef]

129. Guo, T.; Xia, Q.; Shao, Y.; Liu, X.; Wang, Y. Direct deoxygenation of lignin model compounds into aromatic hydrocarbons through hydrogen transfer reaction. *Appl. Catal. A* **2017**, *547*, 30–36. [CrossRef]

130. Barta, K.; Matson, T.D.; Fettig, M.L.; Scott, S.L.; Iretskii, A.V.; Ford, P.C. Catalytic disassembly of an organosolv lignin via hydrogen transfer from supercritical methanol. *Green Chem.* **2010**, *12*, 1640–1647. [CrossRef]

131. Deuss, P.J.; Scott, M.; Tran, F.; Westwood, N.J.; de Vries, J.G.; Barta, K. Aromatic Monomers by in Situ Conversion of Reactive Intermediates in the Acid-Catalyzed Depolymerization of Lignin. *J. Am. Chem. Soc.* **2015**, *137*, 7456–7467. [CrossRef] [PubMed]

132. Galkin, M.V.; Samec, J.S.M. Selective Route to 2-Propenyl Aryls Directly from Wood by a Tandem Organosolv and Palladium-Catalysed Transfer Hydrogenolysis. *ChemSusChem* **2014**, *7*, 2154–2158. [CrossRef] [PubMed]

133. Galkin, M.V.; Smit, A.T.; Subbotina, E.; Artemenko, K.A.; Bergquist, J.; Huijgen, W.J.; Samec, J.S.M. Hydrogen-free catalytic fractionation of woody biomass. *ChemSusChem* **2016**, *9*, 3280–3287. [CrossRef] [PubMed]

134. Kumaniaev, I.; Samec, J.S.M. Valorization of Quercus suber Bark toward Hydrocarbon Bio-Oil and 4-Ethylguaiacol. *ACS Sustain. Chem. Eng.* **2018**, *6*, 5737–5742. [CrossRef]

135. Kumaniaev, I.; Subbotina, E.; Savmarker, J.; Larhed, M.; Galkin, M.V.; Samec, J.S.M. Lignin depolymerization to monophenolic compounds in a flow-through system. *Green Chem.* **2017**, *19*, 5767–5771. [CrossRef]

136. Van den Bosch, S.; Schutyser, W.; Koelewijn, S.-F.; Renders, T.; Courtin, C.M.; Sels, B.F. Tuning the lignin oil OH-content with Ru and Pd catalysts during lignin hydrogenolysis on birch wood. *Chem. Commun.* **2015**, *51*, 13158–13161. [CrossRef] [PubMed]

137. Kim, K.H.; Simmons, B.A.; Singh, S. Catalytic transfer hydrogenolysis of ionic liquid processed biorefinery lignin to phenolic compounds. *Green Chem.* **2017**, *19*, 215–224. [CrossRef]

138. Song, Q.; Wang, F.; Cai, J.; Wang, Y.; Zhang, J.; Yu, W.; Xu, J. Lignin depolymerization (LDP) in alcohol over nickel-based catalysts via a fragmentation–hydrogenolysis process. *Energy Environ. Sci.* **2013**, *6*, 994–1007. [CrossRef]

139. Toledano, A.; Serrano, L.; Pineda, A.; Romero, A.A.; Luque, R.; Labidi, J. Microwave-assisted depolymerisation of organosolv lignin via mild hydrogen-free hydrogenolysis: Catalyst screening. *Appl. Catal. B* **2014**, *145*, 43–55. [CrossRef]

140. Ferrini, P.; Rezende, C.A.; Rinaldi, R. Catalytic Upstream Biorefining through Hydrogen Transfer Reactions: Understanding the Process from the Pulp Perspective. *ChemSusChem* **2016**, *9*, 3171–3180. [CrossRef] [PubMed]

141. Ferrini, P.; Rinaldi, R. Catalytic Biorefining of Plant Biomass to Non-Pyrolytic Lignin Bio-Oil and Carbohydrates through Hydrogen Transfer Reactions. *Angew. Chem. Int. Ed.* **2014**, *53*, 8634–8639. [CrossRef] [PubMed]

142. Molinari, V.; Clavel, G.; Graglia, M.; Antonietti, M.; Esposito, D. Mild Continuous Hydrogenolysis of Kraft Lignin over Titanium Nitride–Nickel Catalyst. *ACS Catal.* **2016**, *6*, 1663–1670. [CrossRef]

143. Zhang, J.-W.; Lu, G.-P.; Cai, C. Self-hydrogen transfer hydrogenolysis of β-O-4 linkages in lignin catalyzed by MIL-100(Fe) supported Pd–Ni BMNPs. *Green Chem.* **2017**, *19*, 4538–4543. [CrossRef]

144. Regmi, Y.N.; Mann, J.K.; McBride, J.R.; Tao, J.; Barnes, C.E.; Labbè, N.; Chmely, S.C. Catalytic transfer hydrogenolysis of organosolv lignin using B-containing FeNi alloyed catalysts. *Catal. Today* **2018**, *302*, 190–195. [CrossRef]

145. Luo, L.; Yang, J.; Yao, G.; Jin, F. Controlling the selectivity to chemicals from catalytic depolymerization of kraft lignin with in-situ H$_2$. *Bioresour. Technol.* **2018**, *264*, 1–6. [CrossRef] [PubMed]

146. Huang, X.; Atay, C.; Korànyi, T.I.; Boot, M.D.; Hensen, E.J.M. Role of Cu–Mg–Al Mixed Oxide Catalysts in Lignin Depolymerization in Supercritical Ethanol. *ACS Catal.* **2015**, *5*, 7359–7370. [CrossRef]

MDPI

St. Alban-Anlage 66

4052 Basel

Switzerland

Tel. +41 61 683 77 34

Fax +41 61 302 89 18

www.mdpi.com

Catalysts Editorial Office

E-mail: catalysts@mdpi.com

www.mdpi.com/journal/catalysts